SCATTERED SKELETONS IN OUR CLOSET

Karen Mutton

SCATTERED SKELETONS IN OUR CLOSET

Adventures Unlimited Press

Scattered Skeletons in Our Closet

ISBN: 1-935487-41-8
ISBN 13: 978-1-935487-41-8

Published by:
Adventures Unlimited Press
One Adventure Place
Kempton, Illinois 60946 USA
auphq@frontiernet.net

www.adventuresunlimitedpress.com

SCATTERED SKELETONS IN OUR CLOSET

ACKNOWLEGEMENT AND PERMISSIONS

I'd like to thank these people:

Glen Kuban for permission to quote from his website and use his photos.

Ed Conrad for permission to quote from his website and reprint images from edconrad.com.

Jim Vanhollebeke for permission to quote from his website and use images from his website http://www.canovan.com/HumanOrigin/

Martin Doutre for permission to quote from his website celticnz.co.nz.

Michael Cremo for permission to quote from his book, 'The Hidden History of the Human Race.'

My family for their love and patience.

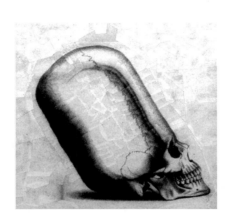

TABLE OF CONTENTS

INTRODUCTION

A war has been quietly raging for 150 years in the hallowed halls of academia that few people are aware of. It is a war between conflicting ideologies for ownership of our extreme past, for accurate information on human origins. The two adversaries are unevenly matched: on one side there is Darwinian evolution with its paradigms which are virtually unchallenged in textbooks and universities around the world. It dovetails neatly with other disciplines like geology, biology, anthropology and genetics and is by far the dominant paradigm about human origins. On the other side is a dedicated group known as creationists who run their own colleges and teach that the Biblical account of creation is absolute and unquestionable. As fundamentalist Christians, they cannot accept Darwinian evolution and do everything to prove that humans did not evolve from apes or that geological eras of millions of years did not exist.

For someone like me, who neither accepts creationism nor aspects of Darwinism such as macroevolution, it is extremely difficult to come across unbiased information. To make matters even more confusing, a more recent theory called Intervention Theory tries to prove that humans were genetically engineered by extraterrestrials. Anomalous skeletal finds, such as those discussed in this book, are often seized by these adherents as evidence of these ancient visitations. Or they are claimed by creationists who are looking for evidence of a global flood. Another fairly recent theory, Hindu creationism, willingly accepts an extreme antiquity for humans because it is supported in the Vedic scriptures, but dispenses with the evolutionary progression of species.

Personally, I am not totally comfortable with any existing paradigm. Creationism and its offspring Intelligent Design exist to defend the Genesis account of creation despite the fact that book was basically a compilation of earlier Mesopotamian myths and legends. However, I don't automatically dismiss all of the theories or findings of trained creationist anthropologists, especially when they reveal shortcomings in Darwinism.

Intervention Theory cannot be proved or disproved. While it is possible that extraterrestrials did visit the earth, or interbreed with humans, there is no conclusive fossil evidence to support this theory. However, it is possible that the 'Starchild' skull might eventually shed some light on this theory when its nuclear DNA is properly sequenced.

The current position of human evolution is far more complicated and convoluted than imagined by early anthropologists, as new finds in Africa, Asia and Europe, as well as mitochondrial DNA testing of fossilized remains, reveals many puzzling and sometimes contradictory insights.

Despite these controversies, Darwinian evolution remains all pervasive in schools and universities. When I was at school in the 1970s a poster of human evolution graced the walls of most classrooms with diagrams showing a gradual progression from ape to humans, via various hairy, hunchbacked ape men. The progression went in a neat linear progression like this- ape- *Australopithecus-Homo habilis- Homo erectus*, Neanderthals and finally the glorious Cro Magnon race. By the time I studied Physical Anthropology at the University of Sydney, the story presented was similar, but there were furiously contested theories about various hominids and dates and many strange offshoots in this family tree were being discovered.

Discoveries over the past few decades have created many conundrums for physical anthropologists as various species are assigned and reassigned, well established fossils are criticized and DNA tests are sometimes shown to be inaccurate. Some well loved species such as *australopithecus* have virtually been dumped from the family tree, while many anthropologists and palaeontologists still argue about what is human and what is simian in the archaeological record, or at what point in time humanity supposedly left Africa to colonise the world.

The purpose of this book is to highlight some of the anomalous findings in the evolutionary paradigm as applied to human physical development. Temporally anomalous discoveries which are often relegated to the basement of museums will be dusted off and given attention. Flaws in the evolutionary sequence will be highlighted. Out of place remains recovered from areas in which they are not supposed to be will also be discussed. This also opens a Pandora's box of information on what is being labelled as suppressed archaeology by such popular writers as Michael Cremo and Lloyd Pye.

Furthermore, chapters will be devoted to legends and actual discoveries of both diminutive humans, like the hobbits of Flores, and giants which were allegedly uncovered in large numbers in the 19th century.

The final chapter concentrates upon some very strange and highly anomalous human skulls which have been recovered over the years. This includes Pye's 'Starchild', the cone heads of ancient Mesoamerica and other malformed skulls.

Rather than giving wholesale support to any particular theory on human origins, my aim is to simply look at the big picture which reveals, through fossils, the wonderful journey of the human race.

PART 1
HUMAN ORIGIN THEORIES

**CLASSIFICATION ISSUES
DATING TECHNIQUES
HUMAN ORIGIN THEORIES
MECHANISMS OF DARWINIAN EVOLUTION
WHAT CREATIONISTS BELIEVE ABOUT
HUMAN ORIGINS
EVOLUTION FAKES AND MISTAKES
CREATIONIST HOAXES AND MISTAKES**

CLASSIFICATION ISSUES

Physical anthropology is the branch of anthropology that studies the evolutionary development of human physical characteristics and the differences in appearance among the peoples of the world. As a science it relies heavily upon classification of species, subspecies and various other physical characteristics. The whole science is inextricably linked to evolutionary theory, and thus any discoveries which don't conform to evolutionary paradigms can be ignored, reclassified, or at worst, totally disregarded. As can be demonstrated in this chapter, the language of this discipline is entirely related to evolutionary theory which influences the whole study of anthropology to the point that any other paradigm exploring human origins is rendered baseless.

TAXONOMY is the practice and science of classification. Taxonomic hierarchies are kinds of things that are arranged in a hierarchical structure. Initially taxonomy referred to the classification of living organisms but now it is applied in a more general sense. **Linnaean taxonomy** is the formal system used in biology in this order: Domain, Kingdom, Phylum, Class, Cohort, Order, Superfamily, Family, Genus and Species. For instance, modern human beings are classified as thus:
Domain –Eukarya, Kingdom- Animalia, Phylum- Chordata, Class- Mammalia.
Cohort- Placentalia, Order-Primates, Superfamily- Hominoidea Family- Hominidae, Genus- Homo, Species- Homo sapiens

PHYLOGENETICS is the study of evolutionary relatedness among various groups or organisms through molecular sequencing data.

CLADISTICS is the hierarchical classification of species based on evolutionary ancestry. Cladistics is different from other taxonomic systems because it focuses on evolution and places heavy emphasis on objective, quantitative analysis. Cladistics diagrams are called cladograms that represent the evolutionary tree of life. Cladistics originated in the work of German entomologist Willi Hennig and the term is often used synonymously with phylogenetics.

HOMINID OR HOMININ? These words have often been used synonymously but there are subtle distinctions based upon Linnaean taxonomy. In the old system, 'hominid' refers solely to the bipedal ape lineage. In the new classification system 'hominid' is a broader system of all the great apes, including fossils like Kenyanthropus.

'Hominin' technically is a subfamily of the Hominidae family. The bipedal apes like chimpanzees, the extinct fossils and living people all are part of the tribe Hominini (hominin). "All hominins are hominids, but not all hominids are hominins," says palaeoanthropologist Lee Berger.
http://news.nationalgeographic.com/news/2001/12/1204_hominin_id_2.html
According to Wikipedia:

- A **hominoid** or ape is a member of the superfamily *Hominoidea*: extant members are the lesser apes (gibbons) and great apes.
- A **hominid** is a member of the family *Hominidae*: all of the great apes.
- A **hominine** is a member of the subfamily *Homininae*: gorillas, chimpanzees, humans (excludes orangutans).
- A **hominin** is a member of the tribe *Hominini*: chimpanzees and humans.
- A **hominan** is a member of the sub-tribe *Hominina*: modern humans and their extinct relatives.

THE HOMINOID LINEAGE includes not only current and extinct apes, but also current and extinct humans in the superfamily of Hominoidea. These hominoids include the genera of homo (humans), Pan (Chimpanzees), Gorilla, Pongo (orangutans) and four genera of gibbons.

It is believed by evolutionists that the initial split between humans and apes was about 15-30 million years ago as various transitory fossils within that time period in Africa and Eurasia have been unearthed. A brief list of these fossilized hominoids or primates includes:

15

EXTINCT AFRICAN APES

Aegyptopithecus is called the Dawn Ape which lived from about 35-33 million years ago in North Africa. This ape is considered to be important because it predates the divergence between hominoids and Old World monkeys and has the reduced teeth and bony eye sockets of simians but no other features which place it in either the monkey or ape group.

Dendroipithecus is known as the 'wood ape' from East Africa and was a fruit eating primate about 22 million years ago. It was similar to Aegyptopithecus but had several features closer to the hominoid lineage such as a rounder, flatter face and more ape-like bones.

Proconsul africanus lived between 14-23 million years ago in the Miocene in Eastern Africa. It had a mixture of Old World monkey and ape characteristics, so is often not placed in the hominoidea superfamily. *Proconsul africanus* was small, weighing about 25 pounds with a brain volume of around 165 millilitres. It has long arm bones which indicate brachiation (arm swinging), although it seems to have been a generalized ape without any special adaptations. *Proconsul nyanzae* was chimpanzee sized and *Proconsul major* was possibly as big as a gorilla.

Morotopithecus was found in Lake Victoria, Africa and was about 20 million years old. It supposedly shows the earliest traces of modern hominoid skeletal features.

EXTINCT EUROPEAN APES

Dryopithecus was originally discovered in France in 1856 and others have been discovered in Austria, Germany and Spain. It was an ape which lived from 12 to 9 million years ago but developed monkey like features of locomotion such as using the flat of its hands rather than knuckle walking which was a characteristic of apes. This ape was larger than most arboreal monkeys and may have regularly come down to the ground. *Dryopithecus* was probably a fruitarian but had strong jaws which enabled it to cope with tougher food such as nuts.

Oreopithecus has been discovered in Italy and lived from 9-7 million years ago. Its skull and teeth are monkey-like but the rest of its skeleton has ape features including no tail. Its long brachiated arms have hands adapted to hanging from branches. Its classification remains controversial in the hominoid evolutionary tree, as some believe it is an Old World monkey, while others think it was an ape.

EXTINCT ASIAN APES

Sivapithecus lived in Pakistan and India from 12.5 to 8.5 million years ago. It had the face, palate and size of an orangutan. *Ramapithecus* was originally found in Nepal in 1932 and initially considered to be an ancestor of humans. With its fairly flat face, parabolic lower jaw and small canine teeth, it was favored as a human ancestor until the mid 1960s. In the 1970s molecular studies indicated a more recent split between humans and apes of 10-7 million years ago and Ramapithecus was demoted as a human ancestor. Similar fossils were uncovered in Turkey, Pakistan and even Fort Ternan, Kenya. Now many researchers believe it is the female form of Sivapitheus and is not a direct ancestor.

Gigantopithecus is an extinct genus of ape which existed from about one million to 300,000 years ago in China, India and Vietnam. *Gigantopithecus blacki* were the largest apes that ever lived standing almost (3 m) 10 feet tall and weighing about 1,200 pounds (540 kg). At first only the teeth of the huge ape had been found in an apothecary shop by anthropologist Ralph von Koenigswald in China, but in recent decades teeth and mandibles have been discovered in China, India and Vietnam. The large, thickly enameled molar teeth and small incisors indicate this giant was a vegetarian and probably resembled a giant orangutan.

Creationist websites often erroneously claim *Gigantopithecus* is a human of gigantic proportions, a relative or *Meganthropus,* but all the evidence indicates it was an ape and not a hominin. Others believe it could still exist in isolated areas as yetis, almas and bigfoot, which is not an impossible scenario considering its fairly recent extinction.

DATING TECHNIQUES

Dating techniques are crucial to the science of physical anthropology as they are used to demonstrate the infallibility of

17

evolutionary theory. Presently there are both absolute and relative dating techniques which have been developed. There are currently more than a dozen techniques for dating the age of rocks and fossils, each with its strengths and shortcomings. The most common are:

STRATIGRAPHY is the branch of geology concerned with the identification, dating and naming of stratified rocks (beds or layers) which may contain fossils. Generally younger rocks and fossils are found near the surface on top of older layers. However, earthquakes, fault lines and folding can tilt, bend or even overturn strata so that dating becomes very difficult. Paleontologists have charted the fossils typical of each layer, allowing them to be able to date most fossils according to other fossils found with them or the type of rock in which they are embedded.

Stratigraphy often relies upon evolutionary models so that fossils found in strata are able to be dated according to where they should lie in the evolutionary sequence, not necessarily their real age. This concept is of vital importance when studying human origins as many hominin fossils have been discovered in strata which may be older or younger than those containing similar fossils, but are dated according to a theoretical construct and not their true age.

Another problem with stratigraphic dating is that of intrusive burials. Some modern humans have been buried in deeper layers, making their age seem much greater. This excuse is always exploited by evolutionists to account for modern human skeletons which have been recovered from great depths, although it is usually possible for a geologist to study the surrounding earth of a burial to check for its age and determine if it is an intrusive burial.

RADIOMETRIC DATING is a form of absolute dating. As rocks form chemical substances which naturally give off radioactivity are encapsulated within them. These turn from one form (or isotope) of an element into another or even into a different element. This is radioactive decay; the original substance is the 'parent' and the new one is the 'daughter.' The parent decay is regular and can be measured by half-life: the time taken for half the number of parent atoms to decay into daughters. Working back from the relative amounts of parent and daughter, it is possible to calculate when the rock was formed.

Absolute dating by radioactive decay uses a series of decay sequences, such as potassium-argon (K-Ar), rubidium-strontium, uranium-thorium-lead and 'carbon dating'. Carbon dating relies on measuring the residual levels of radioactive carbon-14 in the specimen. C14 is most useful for organic matter up to 70,000 years old.

C14 has had its problems and is not entirely reliable. Generally the rate of creation of C14 has been constant, but it can be affected by eruptions of volcanoes or other events which give off large amounts of CO_2 and reduce local concentrations of C14, thus giving inaccurate dates.

Initial, or 'raw' dates need to be calibrated to compensate for the changes in atmospheric C14 and variations in cosmic rays. Comparisons with other dating methods, such as tree growth rings (dendrochronology,) and deep ocean sediment cores, also show the need for recalibration of raw dates.

However, calibration curves can vary significantly from a straight line, so comparison of uncalibrated radiocarbon dates can give misleading results. There are also significant plateaux in the curve such as the one from 11,000 to 10,000 radiocarbon years B.P. which was probably associated with changing ocean currents during the Younger Dryas period.

Once calibrated a radiocarbon date should be expressed in terms of cal B.C, cal A.D. or cal B.P.

Apart from the shortcomings of calibrated vs. uncalibrated C14 dating, it is also only available to date specimens with organic material and cannot be used to date objects beyond 70,000 years.

Potassium-argon or K-Ar dating measures the radioactive decay of potassium which is found in many rocks and minerals. It is based on the fact that some of the radioactive isotope of Potassium, Potassium-40 (K-40) decays to the gas Argon as Argon-40 (AR-40). By comparing the proportion of K-40 to Ar-40 in a sample of volcanic rock, and knowing the slow decay rate of K-40, that rock's date can be ascertained. This dating is the most useful for very ancient geographic and archaeological specimens such as the hominins of Olduvai Gorge and Hadar in Ethiopia.

The limitations of K-Ar dating are:

19

- It works well on any igneous or volcanic rock which hasn't gone through a heating or recrystallization process after its initial formation. Only trained geologists should collect samples in the field.
- Because materials dated using this method are not the direct result of human activity (unlike C14), it is critical that the association between the igneous/volcanic bed being dated and the strata containing human evidence is very carefully established.
- K-Ar dating is accurate from 4.33 billion years to about 100,000 B.P. After that it is inaccurate.

PALEOMAGNETISM is the study of the history of the Earth's magnetic field which has changed many times. As igneous rocks form from molten lava, particles of magnetite line up according to the Earth's magnetic field at the time. Scientists study the direction and degree of particle alignment, allowing them to produce a magnetic chronology. Fossils can be dated from igneous rocks above or below them although the method is only useful for igneous rock containing magnetite.

FLUORINE DATING Bones and teeth fossils absorb natural fluorine from the ground water around them. Rates of fluorine absorption vary greatly but fossils from different levels of the same site, or from comparable sites can be dated relative to each other by analyzing their fluorine content. The more fluorine they hold, the older they are.

ELECTRON-SPIN RESONANCE (ESR) This method measures the resonance or vibration of the electrons of a substance which can reveal the time the substance became stabilized. It measures the resonance under the effect of a powerful magnetic field by microwaves beamed through the specimen.
This technique is used mainly to date minerals but can be useful for dating sedimentary quartz, fossilized teeth, flint and calcium carbonate.

DENDROCHRONOLOGY is the method of dating based upon tree-ring patterns. Generally there is one tree ring per year although this can vary on rare occasions. A benefit of dendro-chronology is that it is based on once-living material and can be accurately dated to a specific year to be used as a calibration and check of radiocarbon dating. The bristlecone pine is one of the oldest living things on Earth and can be used to provide tree ring patterns going back thousands of years, even up to 10,000 years.

Dendrochronology can have its limitations, especially when the wood has been damaged by ants. While useful for tree dating, it is less reliable for dating wooden structures because it only dates the wood itself which may be much older than the structure, or may be a more recent replacement beam in an older structure.

HUMAN ORIGIN THEORIES

Prior to Darwin and his Theory of Evolution, most people assumed that God had created human beings in His own likeness and placed them in the Garden of Eden. This theory has not died out and has enjoyed resurgence in the movements known as Creationism and Intelligent Design. Even more recently, the Vedic story of creation has been used by a couple of scholars to challenge Darwinian evolution.

This chapter gives a brief overview of the main theories which explore the mystery of human origins.

DARWINIAN EVOLUTION

The currently favored human origins paradigm dictates that all animals, plants and humans have evolved from earlier forms and share a common ancestry. Mechanisms of this evolution are both natural selection, where only the fittest survive to hand on their genes, to mutations where beneficial random genetic mutations are passed on to future generations. The Theory of Evolution was first proposed by English naturalist Charles Darwin in the 1850s after he travelled around the world observing many animals and studying fossils. Discoveries in genetics greatly enhanced the theory in the 20th century.

According to this almost universally accepted theory, humans are directly descended from apes which inhabited Africa millions of years ago. Many hominid remains, (apelike humans) apparently fit into this paradigm although there is still a dearth of evidence about so called missing links, transitional fossils displaying mixed traits.

CREATIONISM

This theory which became popular from the 1920s is espoused by fundamentalist Christians primarily in the USA who take the Bible literally. Creationists believe that humans were created by God as related in the book of Genesis and are not related to apes. They generally discount Darwin's theory of evolution and the grand geological timescales favored by scientists. Creationism itself has been split into subgroups such as Creation Science, Young Earth Creationists and Old Earth Creationists. Prominent creationists who refute evolutionary science are Duane Gish and Marvin Lubenow. Lubenow not only has a Master of Theology, but also a Master of Science degree with a major in anthropology.

INTELLIGENT DESIGN

This theory was developed in the 1980s by a group of American creationists who reformulated their theory to circumvent court rulings which prohibit the teaching of creationism as science. They all believe that the designer of the universe and all living things, including humans, is the God of Christianity. Original advocates, mainly from the 'Discovery Institute' included scholars like Phillip Johnson, Michael Behe, Stephen Meyer and William Dembski. The stated purpose of intelligent design is to investigate whether or not existing empirical evidence implies that life on earth must have been designed by an intelligent agent, particularly God.

VEDIC CREATIONISM

This relatively new theory to the human origins debate was originally espoused by Hindu (International Society for Krishna Consciousness, ISKCON) researchers Michael Cremo and Richard Thompson who believe that according to the ancient Hindu scriptures like the Rig Veda and Puranas, human life is millions of years old. Cremo and Thompson's seminal work 'Forbidden Archaeology' catalogued hundreds of anomalous and forgotten reports from the annals of physical anthropology to find evidence of the extreme antiquity of humans.

INTERVENTION THEORY

This recent theory by independent researcher Lloyd Pye agrees to a point with Intelligent Design that Darwinian evolution is incorrect. The core of Intervention Theory is that extraterrestrial beings, known as 'gods' to the ancients, populated the Earth in ancient times with human beings and that hominins are not related to humankind. The original stimulus to this theory is the work of Zecharia Sitchin, who translated the Sumerian tablets and allegedly found reference to extraterrestrial beings which terraformed the planet. Pye came into possession of a highly anomalous Mesoamerican skull which he dubbed the 'Starchild' because he believes it to be of extraterrestrial origin. DNA testing of this skull continues to reveal many anomalous markers in its XY lineage.

An 1871 cartoon of Charles Darwin as an ape.

THE MECHANISMS OF DARWINIAN EVOLUTION

The year 2009 marked the bicentennial of the birth of Charles Darwin, one of the 19th century's most influential scientists. His theory of evolution which totally revolutionised the sciences of biology, geology and anthropology, remains the only theory of human origins acceptable to establishment scientists and historians.

23

SO WHAT EXACTLY DID DARWIN PROPOSE?

Darwin was an English naturalist who sailed around the world for five years on 'the Beagle' collecting plants, animals and fossils. Although apprenticed to be a doctor, he soon preferred natural history and learned the art of taxidermy from a freed slave. One of his earliest mentors was geologist Charles Lyell whose 'Principles of Geology' set out uniformitarian concepts of land slowly rising or falling over immense periods of time. Darwin sided with this belief and rejected catastrophism even though he saw evidence of dramatic earth changes including shells in the Andes and the coast of Chile which had been raised by an earthquake.

Darwin became interested in the theory of 'transmutation of the species' which Lamarck had proposed to describe the altering of one species into another. Another influence came from Malthus's 'An Essay on the Principle of Population' which advocated controlling human population in order to preserve the food supply. Darwin wrote, "In October 1838, that is, fifteen months after I had begun my systematic enquiry, I happened to read for amusement Malthus on Population, and being well prepared to appreciate the struggle for existence which everywhere goes on from long-continued observation of the habits of animals and plants, it at once struck me that under these circumstances favorable variations would tend to be preserved, and unfavorable ones to be destroyed. The result of this would be the formation of new species. Here, then, I had at last got a theory by which to work..." (Darwin, Charles (1958), Barlow, Nora, ed., 'The Autobiography of Charles Darwin *1809–1882'*)

Darwin's theory of 'natural selection' was similar to that independently proposed by naturalist Alfred Russell Wallace and on July 1, 1858 their papers were jointly presented to the Linnean Society. In 1859 Darwin published his seminal 'On the Origin of the Species' which laid out his theory of natural selection, the process by which favorable inherited traits become more common in successive generations of a population of reproducing organisms, and unfavorable traits become less common.

He also put a strong case for 'common descent' whereby a group of organisms is said to have common descent if they have a common ancestor. The controversial term 'evolution' was avoided.

24

In 1871 Darwin published 'The Descent of Man' presenting evidence from numerous sources that humans are animals, an idea which had already been propounded by Thomas Huxley, who showed that anatomically humans are apes, and the anonymously authored book 'Vestiges of the Natural History of Creation' which argued for the evolutionary process. He wrote: "Man with all his noble qualities…with his god-like intellect….still bears in his bodily frame the indelible stamp of his lowly origin."

Darwin was unable to explain the source of heritable variations which would be acted on by natural selection, or how traits passed down from generation to generation. In 1865 Gregor Mendel discovered that traits were inherited in a predictable manner and the science of genetics was born. Early geneticists like Hugo de Vries were critical of Darwin's evolutionary theory, but evolutionary biologists such as J. Haldane, Sewall Wright and Ronald Fisher in the 1920s and 30s set the foundations for the establishment of the field of population genetics. The modern **evolutionary synthesis** is a combination of evolution by natural selection and Mendelian inheritance. The discovery of DNA by Oswald Avery and the publication of its structure in 1953 by James Watson and Francis Crick demonstrated the physical basis for inheritance.

The tenets of the modern evolutionary synthesis are:

1. All evolutionary phenomena can be explained in a way consistent with known genetic mechanisms and the observational evidence of naturalists.

2. Evolution is gradual. Discontinuities amongst species are explained as originating gradually through geographical separation and extinction.

3. Selection is overwhelmingly the main mechanism of change. The object of selection is the phenotype in its surrounding environment. The role of genetic drift is equivocal.

4. The genetic diversity carried in the natural populations is a key factor in evolution. The strength of natural selection in the wild was greater than expected; the effect of ecological factors is important.

5. In palaeontology the ability to explain historical observations by extrapolation from micro to macro-evolution is proposed.

TERMINOLOGY OF EVOLUTIONARY THEORY

- *Natural selection-* a process causing heritable traits that are helpful for survival and reproduction to become more common in a population, and harmful traits to become more rare.
- *Adaptation-* a combination of successive, small random changes in traits and natural selection of those traits best suited for their environment.
- *Genetic Drift-* an independent process that produces random changes in the frequency of traits in a population. This process can culminate in the emergence of new species.
- *Gene flow-* also known as gene migration is the transfer of alleles (member of a pair) of genes from one population to another.
- *Mutations* are changes to the genetic material of an organism. They can be caused by copying errors in the genetic material during cell division, by exposure to radiation, mutagens or viruses.
- *Heredity-* is the process through which evolution in organisms occurs through changes in heritable traits-particular characteristics of an organism. Inherited traits are controlled by genes and the complete set of genes within the organism's genome is its genotype.
- *Phenotype-* The complete set of traits that make up the structure and behaviour of an organism is its phenotype. These traits come from the interaction of its genotype with the environment.
- *Variation-* A substantial part of the variation in phenotypes in a population is caused by the differences between their genotypes. The modern evolutionary synthesis defines evolution as the change over time in this genetic variation.
- *Mechanisms* – for producing evolutionary change are natural selection, genetic Dr.ift and genetic flow.
- *Speciation* – the evolutionary process by which new biological species arise.
- *Macroevolution-* evolution which occurs at or above the level of species, such as extinction and speciation.

- *Microevolution*- smaller evolutionary changes such as adaptations within a species or population.
- *Co-evolution*- interactions between organisms and species where matched sets of adaptations occur.
- *Extinction*- the disappearance of an entire species.
- *Common descent*- the belief that all organisms on Earth are descended from a common ancestor or ancestral gene pool.

HISTORY OF PALEOANTHROPOLGY & ANTHROPOGENESIS

Anthropogenesis is the study of biological evolution which concerns the emergence of *Homo sapiens sapiens* (modern man) as a distinct species from other hominins, apes and placental mammals. The study of human evolution encompasses the scientific disciplines of physical anthropology, primatology. linguistics, anatomy and genetics.

Palaeoanthropology is the study of ancient humans based on fossil evidence, tools and other signs of human habitation. This discipline began with the discovery of a Neanderthal skeleton in 1856 and three years later became wedded to Evolutionary theory as proposed by Darwin in his book 'On the Origin of Species.' Darwin's contemporary, Thomas Huxley illustrated many of the similarities and differences between humans and apes in his 1863 book 'Evidence as to Man's Place in Nature.'

Darwin's theory was eventually accepted by scientists who then surmised that humans share a common ancestor with African great apes, and that fossils of these shared ancestors would be found in Africa.

Dutch anatomist Eugene Dubois, on the other hand, was convinced that the legendary 'missing link', the evolutionary connection between apes and modern humans, could be found in Java, Indonesia. In 1891 he discovered the skullcap of an 'ape-man' while digging into a fossil rich river sediment in Java. He gave the ape-man the name of '*Pithecanthropus erectus*' meaning 'ape-man which stood upright.' Today it belongs to the extinct species known as *Homo erectus*. Unfortunately for him few scientists accepted his finds until the 1930s when German/Dutch palaeontologist Gustav von Koenigswald made similar discoveries in the Dutch East Indies.

In 1924 the first evidence of African hominids was found by Australian anthropologist Raymond Dart who had his students bring him bones from the Taung limestone works, Bechuanaland. He noticed a primate skull which resembled that of a child and discovered the 'Taung' child which he named '*Australopithecus africanus*' meaning southern ape from Africa. It is interesting to note that the Taung child was removed from its original position and was unable to be accurately dated.

Even though Dart had his detractors, including his own mentors Grafton Eliot Smith and Arthur Keith, others, including palaeogeologist Robert Broom, supported Dart and made further australopithecine discoveries in Africa at Sterkfontein and Swartkrans.

It was the work of Kenyan anthropologist Louis Leakey which finally convinced the world that Africa was the 'cradle of mankind.' Working in Olduvai Gorge Tanzania, he not only discovered a new type of Australopithecus which he originally called 'Zinj', but also the skeletons of the first homo species, '*Homo habilis*' the toolmaker.

Meanwhile in China, secluded from the world by the bamboo curtain, impressive discoveries were being made, particularly of *Homo erectus* skeletons.

EVOLUTIONARY THEORIES

Current studies in human evolution concentrate upon when humans or hominins left the African continent to populate the world. The following theories unequivocally accept the premise that Africa is the cradle of the human race.

OUT OF AFRICA THEORY
('EVE' OR 'REPLACEMENT' THEORY)

This dominant theory holds that Africa is the true cradle of mankind where earlier hominins developed into the Homo species over millions of years. Modern *Homo sapiens* evolved, they believe, from a small group of Africans living about 100,000 years ago. This is known as the 'recent African origin of modern humans' (ROA) theory, that *Homo sapiens* evolved in Africa and spread across the world replacing earlier populations of *H. erectus* and Neanderthals.

This theory has gained support by recent research using mitochondrial DNA (mtDNA.) After analyzing genealogies constructed using 133 types of mitochondrial DNA, scientists concluded that all populations were descended from an African woman, dubbed Mitochondrial Eve.

MITOCHONDRIAL EVE HYPOTHESIS

She is the theoretical 'matrilineal most recent common ancestor' (MRCA) for all living humans and is believed to have lived about 140,000 years ago in Ethiopia, Kenya or Tanzania. Geneticists claim that her mitochondrial DNA (mtDNA) is found in every living human through the matrilineal lineage.

This hypothesis was developed by Allan Wilson, Rebecca Cann and Mark Stoneking in the late 1980s. Wilson focused on mitochondrial DNA, genes that sit in the cell but not the nucleus and are only passed from mother to child. By comparing differences in the mtDNA Wilson believed it was possible to estimate the time and place that modern human beings first evolved. According to this theory, modern humans had recently diverged from a single population of Africans while older hominid species like Neanderthal and *Homo erectus* had become extinct without donating to the gene pool.

Of course there were many criticisms of this theory. Only 147 people's DNA were sampled, and of them only two of them were from sub-Saharan Africa. The other 18 'Africans' in the study were actually black Americans. From this tiny sample, it was extrapolated that every human being in the world is a descendant of 'Eve'. Furthermore, geneticist Alan Templeton pointed out statistical and sampling flaws in the study, particularly as its results, he claimed, were in part dictated by the order in which the data were fed into the computer. Other geneticists questioned the reliability of 'molecular clocks' and the rate of mutation in the mtDNA used to calculate Eve.

In the 1990s scientists discovered that mtDNA appeared to

mutate faster than expected which raised troubling questions about the dating of evolutionary events. Using this new calibrated mitochondrial clock, Eve was only dated at 6,000 years ago. A subsequent study that used a far greater area of the mitochondrial sequence did not confirm the high rate of mutation, and dated Mitochondrial Eve at 170,000 (+/-50,000) years.

The Mitochondrial Eve hypothesis dovetails neatly with the Out of Africa theory and is currently favored by the majority of anthropologists. Some misconceptions of the theory are addressed:

- 'Eve' is not the only common ancestor of the human race. All her ancestors and many of her contemporaries are probably also ancestors.
- 'Eve' was not the only female in existence, but is the only one who is a common ancestor of all humans alive today by strictly matrilineal descent.
- 'Eve' didn't necessarily know 'Adam,' the first Y-nuclear male through which his Y chromosomes were passed down intact from father to son.

Y- CHROMOSOMAL ADAM is the most recent common male ancestor through whom all Y chromosomes in living men are descended. By analyzing the Y-chromosome DNA from males around the world, Spencer Wells concluded that all humans today are descended from a single man who lived in Africa around 60,000 years ago.

It should be noted that 'Y-chromosome Adam,' although the so-called male counterpart of Mitochondrial Eve, was not her contemporary.

REGIONAL CONTINUITY THEORY
(MULTI-REGIONALISM THEORY)

This theory favored by anthropologists Milford Wolpoff and Dr. Alan Thorne maintains that there is little supporting evidence that a small group originating in a single geographic region (Africa) replaced the entire population of early humans. They believe in the theory of multi-regional evolution (MTO), that modern humans evolved as a widespread population from existing Homo species such as *H. erectus* which originated in Africa.

OUT OF ASIA HYPOTHESIS

Some earlier palaeontologists contended that Asia was the cradle of mankind and humans evolved from a hominoid known as Ramapithecus. Anthropologists like Dubois and Van Koenigswald conducted successful excavations in both Indonesia and China, finding the first *Homo erectus* fossils ever uncovered. This Out of Asia theory is currently out of favor as anthropologists firmly embrace the Out of Africa model but is tacitly favored by some Chinese anthropologists. However, the fossil record indicates that anthropoid primates lived in this area millions of years ago. It should also be worth noting that the mega volcano Mount Toba on the island of Sumatra exploded catastrophically about 70,000 years ago and almost caused the extinction of the human race. It is difficult to imagine that any hominids in South East Asia escaped this catastrophe.

FAMOUS EVOLUTIONARY ANTHROPOLOGISTS

It is not surprising that all the famous anthropologists and palaeontologists were trained in evolutionary theory, and studied human fossils through a Darwinian lens.

EUGENE DUBOIS (1858-1940) was both an anatomist and anthropologist and was sure that the evolutionary 'missing link' would be found in Asia, particularly in Sumatra and Java. He is credited with discovering the first *Homo erectus* specimen known as Java Man in Indonesia.

ROBERT BROOM (1866-1951) was a South African doctor and palaeontologist, with early expertise in vertebrates and later in hominids. He discovered fragments of six hominids from Sterkfontein which were later classified as *Australopithecus africanus* and is famous for his 1937 discovery *of Paranthropus robustus*.

RAYMOND DART (1893-1988) was an Australian anatomist and anthropologist most famous for his 1924 discovery of the Taung child in South Africa which was later classified as an *Australopithecus africanus*. Upon examining the endocranial cast, Dart immediately pronounced it as the missing link between apes and humans because of its small brain size and human like dentition.

DAVIDSON BLACK (1884-1934) was a Canadian physician and amateur palaeoanthropologist famous for naming *Sinanthropus pekiensis*, now known as *Homo erectus pekiensis* or Peking Man. Possessing a degree in medical science, he showed an interest in human evolution and went to work at Peking Union Medical College. Having acquired two ancient molars from Johan Gunnar Anderson, he received a grant from the Rockefeller Foundation to search for more fossils around Zhoukoudian, China. His team later discovered more teeth and bones at the site but they were unfortunately lost during the war.

FRANZ WEIDENREICH (1873-1948) was a German Jewish anatomist and physical anthropologist who studied human evolution. In 1935 he succeeded Davidson Black as the honorary director of the Cenozoic Research Laboratory of the Geological Survey of China. He studied the Peking Man fossils and originated the Weidenreich theory which postulated that human races evolved independently in the Old World from *Homo erectus* to modern man, while at the same time there was a gene flow between the various populations.

DR. GUSTAV VON KOENIGSWALD (1903-82) was a German palaeontologist and geologist who conducted research on Asian hominins, especially *Homo erectus* in Java. His work led to the discovery of *Homo erectus* skull caps (calvarium) in Sangiran, including that of the controversial *Meganthropus palaeojavanicus*. In later years he studied Asian hominoids such as *Ramapithecus* and concluded that the Hominidae species originated in India.

KENNETH OAKLEY (1911-81) was an English physical anthropologist, palaeontologist and geologist who is best known for his dating of fossils by fluorine content and his exposure of the Piltdown man hoax.

PHILIP TOBIAS (1925-) is a South African palaeoanthropologist and Professor Emeritus at the University of the Witwatersrand, Johannesburg. He is acknowledged as one of the world's leading authorities on human evolution, having excavated at the Sterkfontein caves and most other major southern African sites since 1945. He is most famous for work on the hominids of Olduvai Gorge in Tanzania where, with Louis Leakey, he identified, described and named the new species *Homo habilis*.

LOUIS LEAKEY (1903-72) was the famous Kenyan archaeologist and naturalist who became the patriarch of a family dynasty of anthropologists, including wife Mary, son Richard, daughter in law Maeve and granddaughter Louise. Although a devoted Christian, he was also an evolutionist who pursued a degree in anthropology. Over the next few decades he and Mary excavated at Kanam and Kanjera but their most famous finds were at Olduvai Gorge where they discovered '*Zinjanthropus*' and *Homo habilis*. In 1968 Louis assisted with the formation of The Leakey Foundation, the largest funder of human origins research in the US.

MARY LEAKEY (1913-96) was the second wife of Louis and a famous archaeologist/ anthropologist in her own right. Not only did she assist Louis with his most famous discoveries, but she also discovered the famous Laetoli footprints in the 1960s. Eventually she became director of excavation at Olduvai and trained son Richard in the field.

RICHARD LEAKEY (1944-) accompanied his parents on excavations as a child and became a skilled anthropologist who has worked in Tanzania, Ethiopia and Lake Turkana. Specimens of *Homo erectus, Homo habilis* and *Paranthropus boisei* were uncovered and currently he is a professor of anthropology at Stony Brook University, New York.

FRANCIS CLARK HOWELL (1925-2007) is considered the father of modern palaeo-anthropology. His career included field trips on Neanderthal sites as well as African and Spanish sites. In the Omo

River region of Ethiopia he found fossils of monkeys as well as hominids. Howell also pioneered new dating methods based on potassium-argon radioisotope techniques. He was instrumental in the creation of various evolutionary institutes such as the L.S.B Foundation, Stone Age Institute and the Human Evolution Research Center at UCLA.

DONALD JOHANSON (1943-) is an American palaeoanthropologist who, along with Maurice Raieb and Yves Coppens, discovered the female australopithecine known as Lucy in the Afar Triangle, Ethiopia.

TIM WHITE (1950 -) is an American Palaeoanthropologist and Professor of Integrative Biology at the University of California, Berkeley. In 1974 he worked with Richard Leakey's team at Koobi Fora Kenya and with Mary Leakey and her hominid fossils from Laetoli, Tanzania. He discovered *Ardipithecus ramidus* and *Australopithecus garhi* and collaborated with Johanson on 'Lucy' research. He is director of the Human Evolution Research Center.

MILFORD WOLPOFF (1942-) is an American palaeoanthropologist and the leading proponent of the Multiregional Evolution Hypothesis which attempts to explain human evolution as a consequence of evolutionary processes within a single species. This idea challenges the popular Out of Africa model.

LEE BERGER (1965-) is a palaeoanthropologist, physical anthropologist and archaeologist famous for his research on *Australopithecus africanus* and the 'Taung Bird of Prey' hypothesis. He is also famous for naming the 'Goliath' fossil and announcing the discovery of dwarf humans on the island of Palau. American born, he currently is a professor at the University of Witwatersrand in Johannesburg, South Africa and has a long standing affiliation with the National Geographic Society.

PRIVATE FOUNDATIONS

Without the financial backing of private foundations such as the Carnegie Institution, Rockefeller Foundation and Leakey

Foundation, many excavations which concentrate on the evolution of the human species would never have occurred. The foundations strictly favor the evolutionary model of human origins.

THE CARNEGIE INSTITUTION was founded in by millionaire Andrew Carnegie 1902 in Washington D.C. and divided into 12 departments of scientific investigation, including its pet theories evolution and the big bang universe. John Merriam, president of the Carnegie Institution from 1921-38, believed that science had "contributed very largely to the building of basic philosophies and beliefs," and he gave monetary support to von Koenigswald's fossil hunting expeditions in Java. Merriam was a strong advocate of evolution and was also a trained palaeontologist.

In his 1943 book 'The Garment of God' he wrote: "But there is reason to believe that of concepts in science arising from study of nature, there are none that would be considered to have influenced our belief more deeply than the generalized principles concerning development or evolution, reaching through vast ages in the story of the earth, and leading ultimately to advance in human life and institutions. As an outgrowth of the view of nature seen today...there is strong evidence that this vision of life development affecting us so deeply tends to transmute itself ultimately into emphasis on what we call progress.... As result of this situation one notes that in studying the universe widely in space, and deeply in time, out of our developing experience there tends to grow an attitude toward life that gives perspective instead of formless space, order in the place of aimless movement, confidence in the dependability of the universe and its laws, and faith that the world is so constructed as to maintain the trend of its development or evolutionary progress. Such an attitude towards this world and its meaning is enormously important to us when, as now, complicated dangers and evils seem almost to overwhelm us."
http://www.ucmp.berkeley.edu/about/history/jcmerriam.php
The Carnegie Institute also financed archaeological expeditions to the Yucatan in Mexico.

THE ROCKEFELLER FOUNDATION was founded in 1913 and initially was a charity directed towards Baptist churches and missions but under trustees like J.D Rockefeller, Dr. Simon Flexner and various bank presidents, became more involved in scientific and medical endeavours.

In 1928 the Rockefeller Foundation and other Rockefeller charities became more aligned to five divisions: international health, medical sciences, natural sciences, social sciences and the humanities. Millions of dollars were endowed upon these sciences but there was also the dark side of eugenics research and population control.

The China Medical Board of New York, which promotes health education and research in the medical universities of China, was founded in 1914 as the China Medical Commission of the Rockefeller Foundation. It funded the Peking Union Medical College in 1921 where physician Davidson Black was appointed assistant. Black's main interests were in palaeoanthropology and he undertook various expeditions to find human ancestors.

When Black attended a scientific meeting at which G. Andersson presented the Prince of Sweden with the report of hominin molars found at Zhoukoudian in 1923, he proposed further excavations to be carried out jointly by the Geological Survey of China and the Beijing Union Medical School. Black requested and received funding from the Rockefeller Foundation and undertook excavations which revealed another tooth of '*Sinanthropus*' or China man.

As the grant was to expire in 1929, Black wrote to the directors of the RF and asked them to support further excavations at Zhoukoudian by creating the Cenozoic Research Laboratory. Funds became available and eventually the first skull of *Sinanthropus* was excavated by Pei Wenzhong. Black became a media sensation and further Rockefeller funds were assured.

Michael Cremo gives a more unsettling view of the Foundation's motives. "It thus becomes clear that at the same time the Rockefeller Foundation was channelling funds into human evolution research in China, it was in the process of developing an elaborate plan to fund biological research with a view to developing methods to effectively control human behaviour. Black's research into Beijing man must be seen within this context in order to be properly understood." (Cremo, 'The Hidden History of the Human Race,' p195)

THE LEAKEY FOUNDATION was founded in 1968 to support the 'fieldwork and scientific priorities' of Dr. Louis Leakey.

The Foundation's mission statement is "To increase scientific knowledge, education and public understanding of human origins, evolution, behaviour and survival."

http://www.leakeyfoundation.org/the-foundation/Mission.html

In its first decade the Leakey Foundation provided grants for the field research and discoveries of Louis, Mary and Richard Leakey, Don Johanson, Jane Goodall, Dian Fossey (primatologists) and Birute Galdikas. The Leakey Foundation continues to support the studies of researchers like Zeresenay Alemseged, Jill Pruetz, Dan Lieberman, Frederick Grine, Sileshi Semaw, David Lordkipanidze and many more.

Its website states: "Today the Leakey Foundation awards more than $600,000 in field grants annually for vital new research exploring human evolution. It is the only US funding organization wholly committed to human origins research. Recent priorities have included research into the environments, archaeology and human paleontology of the Miocene, Pliocene, and Pleistocene; into the behavior, morphology and ecology of the great apes and other primate species; and into the behavioral ecology of contemporary hunter-gatherers." (ibid)

Louis and Mary Leakey from Wikipedia.

WHAT CREATIONISTS BELIEVE
ABOUT HUMAN ORIGINS

Creationism in the west is based upon the account of creation according to the Old Testament book of *Genesis*. Since the 1920s it has been a formidable movement across the southern states of the USA with many adherents and schools. Creationists can be split into various groups:

1. Young Earth Creationists believe that the Earth and the universe are thousands rather than billions of years old in accordance with Genesis. They reject the Big Bang cosmology and geological eras. A well known member of this community is Carl Baugh who runs the Creation Evidence Museum in Texas and is a strong proponent of the theory that humans and dinosaurs existed contemporaneously.

2. Old Earth Creationists accept geological evidence of the Earth's extreme age and believe these findings do not contradict the Genesis account but they reject evolution. Subgroups are Gap and Progressive creationism.

3. Creation Science is a creationist movement which attempts to use scientific facts and theories on geology, cosmology and biological evolution to prove the Genesis account of creation.

A key concept is a belief in 'creation *ex nihilo*' that humans and other biological forms were created as unique, unchanging kinds and did not evolve from earlier 'kinds'. Baraminology is a creationist system which classifies life into groups or 'kinds' not related by common ancestry called 'baramins.' These 'kinds' are found in a literal reading of Genesis based upon these passages: Genesis 24: "And God said: 'Let the earth bring forth the living creature after its kind, cattle, and creeping thing, and beast of the earth after its kind.' And it was so. 25: And God made the beast of the earth after its kind, and the cattle after their kind, and everything that creepeth upon the ground after its kind; and God saw that it was good."

In Creation Science fossils are indicative of the Noachian flood which covered the whole world and was described in Genesis. Creation scientists purport to offer a true scientific challenge to Darwinism or Darwinian evolution.

Creation science incorporates catastrophism to account for Earth's geological formations; the belief that cataclysmic events shaped the Earth's surface. Therefore it ignores the scientific principle of Uniformitarianism.

Creation Science became an organised movement during the

1960s and was strongly influenced by Seventh day Adventist George McCready Price who advanced a theory of 'new catastrophism.'

When applied to human origins, Creation science or biology believes:

1. 'Kinds' of plants and animals can only change within fixed limits.
2. Humans and apes have a separate ancestry.

Their arguments against evolution include these themes:

- Missing links or gaps in the fossil record disprove evolution.
- Increased complexity of organisms over time through evolution is not possible due to the law of increasing entropy.
- It is impossible that the mechanism of evolutionary theory is untestable.
- Fossil remains of purported hominid ancestors are not considered to be evidence for a speciation event involving *Homo sapiens*.

According to the Wikipedia article on Creationism, a comparison of major creationist views regarding human origins and biology reveals:

- Young Earth Creationists believe man was created directly by God and that macroevolution does not occur.
- Gap Creationists believe man was created directly by God and macroevolution does not occur.
- Progressive Creationists believe man was created directly by God based on primate anatomy, and creation coexists with evolution although there is no common ancestor.
- Theistic Creationists believe in evolution from primates and evolution of all species from a single ancestor.
- Intelligent Design proponents hold various beliefs and that evolution from primates is possible but divine intervention occurred at some point known as "irreducible complexity."

REFERENCE: **http://en.wikipedia.org/wiki/Creationism**

The most famous creationist anthropologists are Duane Gish, Malcolm Bowden and Marvin Lubenow.

DUANE GISH (1921-) is an American biochemist who was formerly vice-president of the Institute for Creation Research (ICR). He authored several books espousing creationism and his best known work is 'Evolution: The Fossils Say No!' Gish is renowned for his fiery debates with mainstream scientists.

MARVIN LUBENOW is a young earth creationist who has a master's degree in Science with a major in Anthropology. His book 'Bones of Contention' is a creationist assessment of human fossils.

FAMOUS EVOLUTIONARY HOAXES & MISTAKES

PILTDOWN MAN

In 1912 a gravel pit at Piltdown, Sussex in England provided the perfect transitional ape-man fossil when a skull and jawbone were reportedly unearthed. The fossilized remains were of a modern looking skull with an apelike mandible, providing evolutionary anthropologists with their holy grail. This specimen was given the auspicious title *Eoanthopus dawsoni* ('Dawson's dawn-man') after its discoverer Charles Dawson. Accompanied by Arthur Smith Woodward of the geological department of the British Museum, Dawson recovered more fragments, including the lower jaw bone.

Woodward announced that fragments had been reconstructed and the skull was similar to that of modern man except for the small brain size. He claimed that with the exception of two human like molar teeth, the jaw bone was identical to that of a young chimpanzee.

Piltdown man was enthusiastically accepted by such authorities as Sir Arthur Keith who in 1938 unveiled a monument to mark the site of its discovery. Other anthropologists and anatomists were far more cautious. Zoologist Gerrit Miller and French palaeontologist Marcellin Boule immediately thought that the mandible belonged to an ape. In 1923 anatomist Franz Weidenreich examined the remains and reported that they consisted of a modern human skull and an orang-utan jaw with filed-down teeth. Despite these criticisms, Piltdown man remained firmly entrenched in the human evolutionary tree until 1953 when Kenneth Oakley, Le Gros Clark and Joseph Weimer published an article in 'The Times' proving that it was an impudent fake. Piltdown man was composed of a medieval human skull, the 500 year old mandible of an orang-utan and chimpanzee fossil teeth.

Piltdown man was an embarrassment to evolutionary science. Despite its crude construction, it was accepted for several reasons. Firstly, British scientists were happy to accept Piltdown man because of nationalistic pride so that a 'first Briton' could compete against fossil hominids found in European countries like Germany and France. Secondly, Piltdown man displayed the expected large modern brain which scientists believed had preceded the modern omnivorous diet. Thirdly, Piltdown man provided the 'missing link' between apes and humans which proved the superiority of evolutionary science.

The identity of the Piltdown forger remains unknown, but suspects have included Dawson, Arthur Keith, Teilhard de Chardin, Arthur Conan Doyle and numerous others. Charles Dawson is the most likely candidate as he was the initial discoverer of the fossils and already possessed a large antiquarian collection from which he could have accessed the remains.

MISTAKEN IDENTITY- NEBRASKA MAN

A single fossil molar tooth discovered in Nebraska in 1922 allegedly bore characteristics of both humans and apes and came to be known as 'Nebraska man' or *Hesperopithecus haroldcooki*. Many authorities gave support to its discoverer Henry Osborn, director of the American Museum of Natural History and reconstructions, based on this single tooth, were lovingly drawn. When a researcher named William Bryan opposed those biased conclusions based upon a single tooth, he was harshly criticized.

Early 1920s drawing of "Nebraska Man."

Other parts of the skeleton were discovered in 1927, and much to everyone's embarrassment, they belonged to an extinct species of wild American pig called Prosthennops. Nebraska man was, in fact, Nebraska pig!

ORCE MAN

In 1995 'British Archaeology' (No 7, September) announced the discovery of human habitation in Orce, Spain leading back to 1.8 million years ago. In 1982 the British/Spanish team headed by Jose Gibert excavated stone artefacts in upper and lower levels while three bone fragments, believed to be hominid, one of a skull and two from an upper arm were discovered in the upper levels of Orce. One year later they announced that the fragment belonged to a human child, making it the oldest human discovered in Europe.

Later to the embarrassment of many, Orce man turned out to be most likely the skull cap of a 6 month old donkey. No wonder it quietly disappeared from the history book.

CREATIONIST HOAXES AND MISTAKES

MOAB MAN

In 1971 two skeletons bearing a greenish tinge were unearthed by amateur geologist Lin Ottinger in the Big Indian Copper Mine near Moab, Utah. A bulldozer had disturbed the 15 feet of ground above the skeletons and damaged them. Archaeologist John Marwitt of the Utah Statewide Archaeological Survey excavated the bones and described them as resting in loose ground, unfossilized and stained green from copper sediments. He concluded that they were intrusive burials a few hundred years old.

A local reporter from the 'Times Independent' interviewed Marwitt and then wrote an article presenting the find as a geological mystery. This angle caught the attention of creationist Clifford Burdick who classified the finds as out-of-place fossils in a creationist journal. In the mid 1980s another creationist, Carl Baugh, purchased

Moab Man.

one of the skeletons from Lin Ottinger and displayed it at his 'Creation Evidence Museum' in Glen Rose, Texas.

In the late 1980s some of the bones were dated by a UCLA lab to be 210 +/- 70 years. Later carbon tests from other organic materials in the mine returned an earlier date of about 1450 +/- 90 years, although it is likely that the bones were a more recent intrusive burial.

MALACHITE MAN

Creationists Don Patton and David Willis were claiming in the late 1990s that green bones recently excavated in Utah represented humans from the Cretaceous era. These new bones were dubbed 'Malachite Man' but on their website the photos were actually from the original Moab man excavations. **http://www.bible.ca/tracks/ malachite-man.htm**

The current 'Official World Site of Malachite Man' does show images from two distinct excavations in the 1970s and 1990s which yielded greenish bones.

The additional excavations which took place at the same site with Patton in attendance in the 1990s revealed the burials were apparently older.

Malachite Man.

However, more recent excavations by Julie Howard and other archaeologists from the Bureau of Land Management indicate that the greenish, unfossilized bones are probably from a similar age as Moab man.

Patton's website on Malachite man makes numerous claims which were not substantiated by archaeologists such as:

- The bones were found in Cretaceous deposits, making them millions of years old. This he does by claiming the bones are in the same Cretaceous Dakota sandstone as the dinosaurs bones at Dinosaur National Monument, despite the fact the latter is 50 million years older and a Jurassic deposit.
- The bones were found 50 feet below the surface. The original

bones were discovered at 15 feet and not 50 feet, although the 1990s finds could have been deeper as they were older.

- He claims some of the bones are articulated, which supports his claim of a global flood. This is a weak argument as articulated skeletons can be found in other burials.

A mandible, purportedly found at the same site, contains teeth composed entirely of turquoise according to Patton. He includes a colored photo of this mandible although its color could have easily been altered by Photoshop. Considering that the bones were unfossilized, this claim is dubious.

Despite Patton's claims, it is more likely that the bones of Moab/Malachite man represent a number of intrusive burials or accidental entombments of native Americans in a mining environment. The bones are modern and unfossilized and the only anomaly is their green mineral color.

REFERENCES and IMAGES:

'Moab Man' – 'Malachite Man,' G. Kuban **http://paleo.cc/ paluxy/moab-man.htm**

'Official World Site Malachite Man,' **http://www.bible.ca/tracks/ malachite-man.htm**

CREATIONIST FOOTPRINTS

Strict (usually young earth) creationists have for many years made claims of human or giant tracks occurring alongside fossilized dinosaur tracks. The implications of these claims are clear- both modern geology and human evolution are falsehoods. Of course it follows that the Biblical creation with its great flood story is correct.

This chapter examines some of these creationist claims and whether there is any corroborating scientific proof.

PALUXY PRINTS The Paluxy River near Glen Rose in Texas has yielded many dinosaur tracks and, some creationists claim,

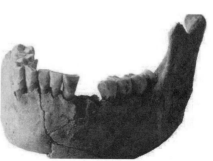
The so-called Turquoise Jaw.

ancient human tracks contemporaneous with the dinosaur tracks. Because they were found together, the young earth creationists claim that both dinosaurs and humans made the tracks while running from the rising waters of the great flood. It is important to realize that young earth creationists do not believe that the dinosaurs lived over 60 million years ago, and instead believe in the Genesis account of creation.

To discuss these prints, two questions need to be addressed:
1. Are they human prints?
2. If they are human, were they contemporaneous with the dinosaur prints?

In the 1980s Glen Kuban, a biology graduate from the College of Wooster in Ohio, studied the Paluxy tracks "while leaning towards strict creationism and hoping to find and better document the supposed human tracks."

His findings were: "We did not find any genuine human tracks in the Paluxy, but did find that the markings in question were due to a variety of phenomena, the most common being elongate, metatarsal dinosaur track, whose digits were often subdued by erosion, mud collapse, and/or infilling....Other alleged human tracks were found to be erosional features, ambiguous marks of uncertain origin, and in some cases, doctored markings or outright cargings (most of the latter on loose blocks of rock.)"

"My Paluxy work helped prompt many creationists to largely abandon the 'man track' claims. However, I would like to clarify that my Paluxy work was not done to attack creationism or Christianity, but to help set the record straight on the true nature of the Paluxy evidence."

http://paleo.cc/paluxy/gkbio.htm

Kuban examined each of the Paluxy sites and made this assessment:

1 The Taylor site- "The most thorough analyzes of the alleged human tracks here are elongate, metatarsal dinosaur tracks...made by dinosaurs that, at least at times impressed their soles and heels as they walked. When the digit marks of such tracks are subdued by one or more factors (erosion, sediment infillng, or mud-collapse.) They often resemble giant prints...When the tracksite surface is well-cleaned, at least some of the tracks in each trail show shallow

tridactyl (three-toed) digit impressions indicating dinosaurian origin....Claims in the 1990s by Carl Baugh and associates that some of these tracks have human prints within them or overlapping them have been shown to be as baseless as the original claims."

2. State Park Ledge – "Included were some alleged striding trackways, child prints, and even a supposed bear print. However, careful analysis of the supposed prints here indicates that they are merely natural irregularities and erosional features of the substrate. None show a clear and natural suite of human features (especially in regards to bottom contours,) and the alleged striding trails do not show consistent human-like stride patterns. Many past 'man track' advocates had applied water, oil and other substances to the markings to encourage the appearance of human shapes; however without selective highlighting none show clear human features."

3. The Baugh/McFall Sites were originally considered human but "later acknowledged by other creationists to consist of eroded, elongate dinosaur tracks."

Kugan concluded: "Although genuine dinosaur tracks are abundant in Texas, claims of human tracks have not withstood close scientific scrutiny, and in recent years have been largely abandoned even by most creationists."

http://paleo.cc/paluxy/mantrack.htm

THE BURDICK PRINT is a very clear impression of a giant foot allegedly dug up from Glen Rose in Texas. However, it is on loose blocks and shows serious anatomic errors which indicate it was carved in recent times.

In the 1920s and 30s Glen Rose resident Wayland Adams, and possibly more people, carved tracks into slabs of stone to sell to tourists. Years later he described his technique to a group of creationists.

Two views of the Burdick Print.

Carl Burdick with his stone footprints.

1. After finding a suitable-sized stone slab, he would carve a footprint using a hammer and chisel.
2. A center punch was used to simulate raindrops, followed by an application of muriatic acid to tone down the chisel and punch marks.
3. After immersing it in manure for a few days, the edges of the slab were chipped to give the impression of a track chiselled from the riverbed.

The so-called Burdick print on a limestone slab is about 18 inches long by 13 inches wide and about 5 inches thick. Anatomically, apart from being of gigantic proportions, the print is too wide at the ball and too narrow at the heel and the toe depressions are far too long. A thorough description of the anatomical mistakes can be found at **http://paleo.cc/paluxy/wilker6.htm**

Needless to say, the Burdick prints are still embraced as natural by hard core creationists such as Don Patton and Carl Baugh but do not warrant any further description in this book.

ALVIS DELK PRINT Carl Baugh announced on his Creation Evidence Museum website in 2008 that he was in possession of "a pristine human footprint intruded by a dinosaur footprint" discovered also at Glen Rose. Baugh had his footprints scanned at a laboratory where 800 X-rays and a CT scan "verified compression and distribution features clearly seen in both prints, human and dinosaur. This removes and possibility that the prints were carved or altered."

http://www.creationevidence.org/index.php?option=com_content&task=view&id=48

The Alvis Delk Print.

The image is indeed impressive but has attracted a lot of criticism. According to Kuban, the dinosaur print was discovered in 2000 by 'amateur archaeologist' and creationist Alvis Delk, but the footprint was only observed in 2008 when the slab was cleaned. There is no in situ documentation, such as the exact location of the original track bed or even if it is from Glen Rose. Anatomically the human foot is also problematic, especially the toes and the depth of the impression. The 'intruded dinosaur track,' identified as Acrocanthosaurus by Baugh, also has strange dimensions, leading many to suspect it is a clever carving.

He concluded, "The Alvis Delk Print is not a convincing human footprint in ancient rock....The uncertain circumstances of the discovery, lack of in situ documentation, knowledge that similar tracks have been carved in the Glen Rose area, the serious morphological abnormalities in the prints and the considerations about potentially misleading scanning artifacts such as beam hardening, point to the strong likelihood that both the "human footprint" and dinosaur track on this loose slab were carved or heavily altered from less distinct depressions."

http://paleo.cc/paluxy/delk.htm

THE 'CRETACEOUS FINGER'

This elongate rock is claimed as a fossilized finger by Carl Baugh who has had it CAT scanned, allegedly revealing human bones. According to Baugh it was discovered in a pile of loose gravel north of Glen Rose, although there is no corroborating evidence that it was an in situ discovery in Cretaceous rocks.

The "Cretaceous Finger."

Apart from its dubious provenance, anatomically there are a number of inconsistencies such as the shape of the finger nail, lack of knuckles and lack of any trauma at the proximal end. Yet Baugh claims that a spiral CAT scan reveals "replaced bone, tissue and ligaments. It has been identified as the fourth finger on a girl's left hand." What the scan clearly shows is some darker areas toward the center of the object which is normal in any natural stone.

Many creationists have also rejected the stone as a human digit, and Kuban concludes, "Baugh and other promoters of the 'fossilized finger' have not conclusively established that it is a real fossil. Nor have they demonstrated a clear association with an ancient formation, undermining its possible value as an out-of-place object. Without this evidence, the object is no more than a curiosity, not a reliable out-of-place fossil."

http://paleo.cc/paluxy/wilker6.htm
REFERENCES: 'The Paluxy Dinosaur/Man Track Controversy,' Glen J. Kuban **http://paleo.cc/paluxy/paluxy.htm**
Creation Evidence Museum
http://www.creationevidence.org/index.php?option=com_content&task=view&id=48

PART 2
THE TANGLED TREE
OF EVOLUTION

ACCORDING TO DARWINIAN EVOLUTION
THE AUSTRALOPITHECINE DEBATE
HOMO HABILIS
HOMO ERECTUS
ARCHAIC HOMO SAPIENS
RECENTLY DISCOVERED FOSSILS

Until quite recently human evolutionists portrayed a fairly lineal rate of descent, indicating that one species evolved into another. Discoveries over the past few decades have revealed that the evolutionary tree has many branches and that some hominin species coexisted for millennia. Furthermore, even classification and the very existence of various hominin species is questioned by some anthropologists. To make matters even more complicated, modern looking *Homo sapiens* have allegedly been discovered in very ancient layers. These discoveries are discounted as intrusive burials or fakes by anthropologists and seized upon by creationists as evidence that humans existed in the early geology of this planet.

ACCORDING TO DARWINIAN EVOLUTION

In the early 20th century various evolutionary anthropologists were postulating that the earliest forms of humans would be uncovered in Africa because of its great apes and primitive tribes. A few dissidents such as Dubois and Black were placing Asia as the cradle of mankind, especially Indonesia with its orang-utans and rich fossil beds.

In the 19th century Neanderthal fossils had been unearthed in Europe, but they were not considered to be primitive enough, so it was to Africa that scientists turned to seek the earliest humans. In 1925 the first pre-Neanderthal skeleton, the Taung skull, was discovered, although large scale excavations were not carried out in Africa for several more decades. The discovery of the Taung child and many subsequent remains known as australopithecines, led to controversies that are still creating debate in anthropological circles.

THE GREAT AUSTRALOPITHECUS DEBATE

Despite decades of discoveries, including skeletal remains named Lucy, Mrs Ples, Nutcracker Man etc, numerous controversies continue to be debated about these African 'southern apes' amongst anthropologists and creationists such as:

* Were the australopithecines apes, the earliest hominins or an

evolutionary 'missing link between the two species?
- Were the australopithecines direct human ancestors or an evolutionary dead end?
- Were they just another extinct species of ape?
- Were they the first true bipeds?

SOUTH AFRICAN AUSTRALOPITHECINES

The original australopithecine, the 'Taung Baby', was discovered in 1925 by Australian Professor Dart of the University of Witwatersrand in Johannesburg, South Africa. It had been blasted from a lime quarry near Taung and sent to the professor in a box of animal skeletal remains. Dart cleaned away the rock to reveal the remains of an almost complete face and mandible, with an endocast of the braincase. He noted that the skull was unexpectedly large for an ape, encasing a brain of about 500 cubic centimeters, and lacked bony ridges above the eyes or protruding jaws. The canine milk teeth were small and the teeth were human-like including the pattern of cusps of each back tooth. The foramen magnum, the spinal cord opening, was set toward the center of the skull as in humans, rather than toward the rear as in apes. Even though Dart had no real way of knowing the fossil's age because the site had not been excavated properly, nor had the fossil been examined in situ, he estimated its age to be one million years because of its primitive morphology. Dart named his Taung baby *Australopithecus africanus*- the great African southern ape.

Many anthropologists were more cautious, including Grafton Elliott Smith, Sir Arthur Keith and Sir Arthur Smith Woodward. Elliott Smith wrote, "It is unfortunate that Dart had no access to skulls of infant chimpanzees, gorillas, or orangs of an age corresponding to that of the Taung skull, for had such material been available he would have realized that the posture and poise of

Researcher holding the "Taung Baby" skull.

53

the head, the shape of the jaws, and many details of the nose, face and cranium upon which he relief for proof of his contention that Australopithecus was nearly akin to man, were essentially identical with the conditions met in the infant gorilla and chimpanzee." (Cremo, p246)

The young age of one million years confused anthropologists because it did not fit the current evolutionary paradigm which decreed that early human ancestors should have a large brain and primitive apelike jaw. Dart's report was supposed to be impartial but he emphasised the Taung Baby's human like features while diminishing the apelike ones. Other anthropologists opposed the idea that the first humans evolved in Africa, as Asia seemed a more 'noble' cradle of mankind.

Dart's Taung child was not accepted by anthropologist for decades. One of his only supporters was Dr. Robert Bloom, a medical doctor and world expert on the remains of mammal-like reptiles. They separated the jaws of the 'ape-child' in 1929 and revealed teeth which were human like and tentatively dated the child at around 2 million years old.

Broom became Curator of Vertebrate Palaeontology and Physical Anthropology at the Transvaal Museum, Pretoria, and began excavations at Sterkfontein. He soon discovered a cranial endocast, skull base, upper jaw and brain case fragments of another australopithecine creature that he called '*Plesianthropus transvaalensis*', nicknamed Mrs. Ples and dated to 2.15 mya. However, this fossil bore little resemblance to Dart's baby, and had more pronounced apelike features.

Yet another fossil discovered by Broom showed very different characteristics from the Taung child or Mrs. Ples. In 1948 Broom, with the help of a schoolboy, discovered a lower jaw, some teeth and a skull fragment from Kromdraai, near Sterkfontein, which was

Mrs. Ples.

approximately 800,000 years old. These were of a hominid with a heavier build and more powerful jaws and teeth which became known as *Paranthropus robustus* or 'robust near-man'.

However, a femur discovered at Sterkfontein called TM1513 appeared to be essentially human, although it should be noted that it was only assumed to belong to an australopithecine. A strong advocate for a bipedal australopithecine, Dr. Robert Broom found more skeletons at Swartkrans and Sterkfontein, including parts of the pelvis, legs and spinal column which convinced him that the creatures were bipedal.

EAST AFRICA

Olduvai Gorge in Tanzania is part of the Rift Valley system. Human fossils were originally recovered there by Hans Reck in 1913 and later by famous palaeontologists Louis Leakey and his wife Mary. In 1959 the Leakeys discovered another skull which was thick boned and heavily built. It resembled *Australopithecus robustus* but had larger teeth, so Leakey called it a new species *Zinjanthropus boisei*, after Zinj, an Arabic name for East Africa and Charles Boise, his sponsor. This creature was known affectionately as Zinj, or 'Nutcracker Man' because of its enormous molar teeth and jaw.

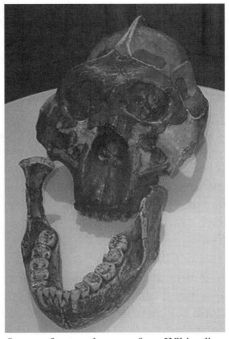

Stratigraphic dating initially yielded an age of around 600,000 years, but later it was dated by radiocarbon dating to 1.8 million years old. Leakey found stone tools nearby and called the hominid the first 'true man'. However, despite his large grant from the National Geographic, lavish magazine articles replete with pictures of the creature in a family setting, Leakey's specimen was

Image of nutcracker man from Wikipedia.

55

quickly demoted to *Australopithecus robustus*. Its huge apelike jaw, flattened cheeks, massive orbital ridges and sagittal crest were far less manlike than earlier specimens, which presented a problem for the evolutionary anthropologists.

Currently Nutcracker man is not considered a human ancestor, but *Paranthropus robustus*, a hominid which split from the family tree about 2.5 million years ago. Other robust specimens have been dated from 2.5 to 1 million year ago, making them contemporaries of other hominids such as *Homo habilis* and *Homo erectus.*

ETHIOPIA

In the 1970s attention shifted from East Africa to the Afar region of Ethiopia which had yielded fossils that were up to four million years old. An international team comprising (among others) French anthropologist Yves Coppens, Geologist Maurice Taieb and American anthropologist Donald Johanson, had uncovered thousands of fossils, including four hominid femur fragments. Dated at over 3 million years old, they revealed a knee joint apparently adapted for upright walking. In 1974 various teeth and mandibles showing both human and ape features were uncovered, leading to the official statement: "We have…extended our knowledge of the genus Homo by nearly 1.5 million years. All previous theories of the origins of the lineage which leads to modern man must now be totally revised." ('The Dawn of Man,' Steve Parker, p88)

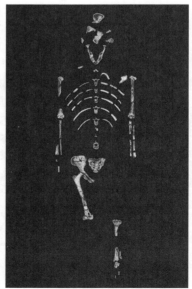

Many anthropologists reserved judgement as the fossils showed some australopithecine traits and some human ones without any indication of brain capacity. Were they australopithecine or Homo fossils?

On Christmas Eve 1974, Donald Johanson and colleague Gray found a hominid arm bone jutting out of a slope at Hadar. More bones revealed a partial female hominid skeleton

The skeleton of "Lucy."

56

catalogued as AL 288-1 Partial Skeleton, or *Australopithecus afarensis* 'Southern ape from Afar.' Its popular name was 'Lucy.' Lucy was nearly 4 million years old and had a small brain but appeared to be bipedal.

About 40% of the skeleton was eventually recovered revealing that the female hominin was only 1.1 m (3'8") tall and weighed only about 29 kg- the stature of a chimpanzee. However she had a pelvis and knee bones which appeared, essentially modern.

In 1975 Johanson and his colleagues discovered more remains they called the 'First Family' consisting of about 13 hominids dating from more than 3 million years ago. They were larger than Lucy, with a combination of modern and apelike features.

"Lucy" reconstructed.

An excavation in Dikika, near where Lucy was recovered, found an entire skull and torso of a 3 year old girl named 'Selam' or 'Baby Lucy' in 2000. Designated *Australopithecus afarensis*, this find comprised a skull, torso and limbs. This creature was adapted to bipedalism and arborealism and revealed features similar to those of Lucy.

According to Wikipedia, the current australopithecine lineage is:

Five main species *A.amanesis, A. afarensis, A. africanus and A. robustus A. boisei*- all found in Africa and minor

Selam.

57

species, *A. garhi, A. bahrelghazali and Paranthopus aethiopicus.*

- *Australopithecus anamensis* is the oldest of the line. The first one was discovered in 1976 at Kanapoi in Kenya and is about four million years old. The 'Kanapoi Hominid' consists mainly of a lower left humerus. Another bone, a tibia belonging to a hominid 4.1 million years old is the oldest evidence for hominid bipedalism.

- An *A. afarensis* skeleton unearthed in Hadar Ethiopia in 1974 is estimated to be about 3.4 million years old. The bones consist of portions of both legs, including a complete knee joint which closely resembles a miniature human knee.

'Lucy' is an incomplete skeleton of an adult female discovered at Hadar in 1974. She was about 107 cm (3'6") tall and weighed about 28 kg (62 lbs) in weight. Her pelvis, femur and tibia show that she was a biped.

A group of about 13 hominids was discovered by Donald Johanson's team at Hadar in 1975. Although these remains are of various sized specimens, they are dated to 3.2 million years, the same age as Lucy.

The Laetoli footprints discovered in 1978 at Laetoli in Tanzania are estimated to be 3.7 million years old. The trail across volcanic ash consists of the fossilized footprints of two or three bipeds ranging from 4 feet tall 120cm) to 4'8" (140cm). These prints have been assigned to *A. Afarensis.*

- *A. bahrelghazali* is a disputed hominid from Chad dated at about 3.6 million years. It apparently has similar features to afarensis but hasn't been widely studied and many anthropologists do not believe it is a separate species.

- *Australopithecus africanus* existed between three to two million years ago and is known as 'gracile,' It was bipedal with a larger body and brain than afarensis. The teeth are much larger than those of humans but more similar to humans than to apes. The Taung Child, discovered by Raymond Dart in South Africa in 1924 had a brain size of 410 ccm (440ccm for an adult) and small canine teeth.

Further *A. africanus* fossils were discovered by Robert Broom in Sterkfontein, South Africa in 1947. One of these skeletons yielded a complete vertebral column, pelvis, rib fragments and part of a femur. The humanlike pelvis suggests africanus was bipedal.

- *A. garhi* is a gracile dating from 2.5 to 2.6 million years found in Ethiopia. It may have been a tool maker and was generally considered to have more advanced features than either africanus or robustus.
- *Paranthropus aethiopicus* is the disputed designation for a primitive 'robust' skull from Omo in Ethiopia and the 'black skull' from Turkana in Kenya. It is about 2.5 million years old.
- *A. robustus* lived between 1.5 and 2 million years ago. With a similar body to africanus, its skull and teeth were more robust. The large face is flat with no forehead and large brow ridges. Its lower jaw teeth were massive for grinding whereas its front teeth were relatively small. Most robustus skeletons are found in South Africa.
- *A. boisei* was similar to robustus but had even larger teeth and a wider face. It lived between 2.1 and 1.1 million years ago. One of the most famous specimens discovered was the 'Nutcracker Man' by Mary Leakey in Tanzania (1959). The cranial capacity of this species is about 500ccm to 550ccm.

CONTROVERSIES SURROUNDING AUSTRALOPITHECINES

These controversies include the position of australopithecines in the hominid family tree, species classification and the incidence of bipedalism. They remain open to interpretation by various anthropologists and are thus 'fluid.'

Man or ape?

English scientist Sir Arthur Keith believed it was a primitive ape. Louis and son Richard Leakey believed that *Australopithecus* was a very early, very apelike offshoot from the main evolutionary line.

In the early 1950s Sir Solly Zuckerman published extensive biometric studies showing *Australopithecus* is much more apelike than humanlike. From the 1960s to 90s Charles Oxnard decided that the brain, teeth and skull of the australopithecines were like those of apes. The shoulder bone appeared to be adapted for tree swinging while the hand bones were shaped like an orang-utan's. Most damningly, the pelvis was adapted for quadupedal locomotion, as was the femur and ankle structure.

According to Johanson, Lucy's discoverer, *Australopithecus afarensis* had "smallish, essentially human bodies." However, when Tim White reconstructed the skull using fragments from other First Family individuals, Johanson felt it "looked very much like a small female gorilla." Other critics like Susman, Stern and Oxnard challenged Lucy's bipedal status by declaring her shoulder blade was almost identical to an ape's. The arm and leg bones were like those of tree climbing primates, as were the long curved finger bones.

COMPARATIVE DATING ISSUES

According to evolutionary theory, older species are supposed to be more 'primitive' than more recent species. If humans evolved from apes, the earlier fossils should show more apelike characteristics than later fossils, which is usually the case. However, there are exceptions.

On a general level, the robust species, often designated as *Paranthropus,* are dated later than the gracile species and display more primitive, apelike characteristics such as large grinding teeth, flaring cheeks, prognathism and sagittal crests.

'The Black Skull' was discovered west of Lake Turkana by Alan Walker of John Hopkins University in 1985. Designated KNM-WT 17000, it was similar to an *Australopithecus boisei,* but was 2.5 million years old, older than the most ancient robust australopithecines which supposedly preceded it. This primitive looking skull was given the designation of *Paranthropus aethiopicus* and shared many traits with *afarensis* which may have been a direct ancestor.

This skull had a sagittal crest and zygomatic arch adapted for heavy chewing as can be seen with gorillas. Its face was prognathic and its brain capacity small at 410 ccm. Most anthropologists believe that this primitive hominid was from a lineage which had already split from the later Homo line.

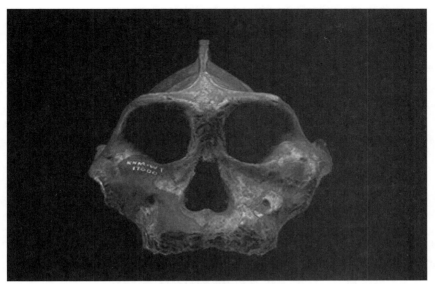

KNW-WT 17000.

Bipedalism is one of the most hotly contested characteristics of the australopithecines by creationists who often accuse evolutionary scientists of fraud.

Adapting an upright posture and bipedal gate, features unknown in the ape family, was associated with great changes in the hominid skeleton. A principle adaptation was the change in position of the foramen magnum, the opening through which the spinal cord enters. In humans it is situated at the base of the skull whereas in apes it faces more towards the rear. The pelvis adapts to bipedalism, being straight, narrow and short. In humans the pelvis is broad, giving stability and allowing wide birth canal in females for the large brained baby.

In humans the backbone rises vertically above the legs and the knee bone, is able to straighten fully and is adapted to withstand upright stresses. The ankle is large and strong with firm joints.

Bipedalism has always been a defining characteristic of humans in the primate family. Evolutionary anthropologists once believed that an enlarged brain was the first human characteristic to evolve, followed by smaller teeth and jaws and finally bipedalism. The discovery of the Taung child and other australopithecines overturned this theory, as they had small brains but exhibited characteristics of bipedalism.

61

The Taung child, with its foramen magnum positioned beneath the skull, rather than towards the back like an ape, was probably a biped. The pelvis discovered in 1946 by Broom at Sterkfontein also was that of a biped. 'Lucy' was a partial skeleton with the pelvis and a femur adapted to bipedalism. According to Wikipedia; "Johanson was able to recover Lucy's left innominate bone and sacrum. Though the sacrum was remarkably well preserved, the innominate was distorted, leading to two different reconstructions. The first reconstruction had little iliac flare and virtually no anterior wrap, creating an ilium that greatly resembled that of an ape. However, this reconstruction proved to be faulty, as the superior pubic rami would not have been able to connect if the right ilium was identical to the left. A later reconstruction by Tim White showed a broad iliac flare and a definite anterior wrap, indicating that Lucy had an unusually broad inner acetabular distance and unusually long superior pubic rami. Her pubic arch was over 90 degrees, similar to modern human females. Her acetabulum, however, was small and primitive." **http:// en.wikipedia.org/wiki/Lucy_(Australopithecus)**

These two reconstructions form the basis for some of the creationist criticisms of a bipedal Lucy. However, creationists maintain consistently that Lucy and the australopithecines were merely apes as there were no 'transitionary' fossils.

A more valid creationist argument is that the leg bones attributed to Lucy belonged to modern humans. Donald Johanson initially found a heel bone and eventually recovered the femur in another location 60 or 70 meters higher in the strata and a few kilometers away. This should ring certain alarm bells and the discrepancy was tackled by none other than the 'Scientific American' in Aug. 2005 by authors William E.H. Harcourt-Smith (American Museum of Natural History) and Charles E. Hilton (Western Michigan University). They claim that the fossil reconstruction of Lucy is based on a mixture of fossils, some from the 3.2 mya *A.afarensis* collection and some from the 1.8 mya *Homo habilis* collection. They claim that one of the bones, the navicular, used to determine that the *A. afarensis* foot was arched, actually was a *Homo habilis* fossil foot bone, not an *A. afarensis* fossil foot bone.

After studying the naviculars of apes, *A. afarensis* and *H. habilis*, Harcourt-Smith and Hilton claimed, "*A. afarensis* almost certainly did not walk like us or, by extension, like the hominids at Laetoli." (Wong, K., Footprints to fill, *Scientific American*, pp.18–19, August 2005)

Creationist Lubenow made these valid points in his article 'Was Lucy Bipedal?' "In over twenty years, to my knowledge, no evolutionist until now has mentioned that the relationship between the *A. afarensis* foot, the Laetoli Footprints, and thus the 'proof' of *A. afarensis* bipedality was based upon a faulty reconstruction. It is hard for me, an outsider, to know whether or not this is a case of outright deception. Was this matter known, but not mentioned, by palaeoanthropologists generally? Or was it known by only a very few who were involved in the reconstruction of Lucy's foot? It is certainly a striking example of the failure of evolutionists to inform the public regarding the actual state of the evidence in the most important 'alleged event' of human evolution." http://creation.com/ was-lucy-bipedal-journal-of-creation-tj

Vedic creationist Michael Cremo also believes that Lucy and the australopithecines are extinct apes. He wrote, "As for the body of *A. afarensis*, Randall L. Suman, Jack T. Stern, Charles E. Oxnard and others have found it very apelike, thus challenging Johanson's view that Lucy walked upright on the ground in human fashion. Lucy's shoulder blade was almost identical to that of an ape. The shoulder joint turned upward, indicating that Lucy's arms were probably used for climbing in trees and perhaps suspending the body. The bones of the arm were like those of tree-climbing primates, and the spinal column featured points of attachment for very powerful shoulder and back muscles. ...The hip and leg bones were also adapted for climbing, and the foot had curved toes that would be useful in grasping branches of trees." (Cremo p261)

Laetoli footprints These amazing small human footprints were discovered in volcanic ash in Tanzania which is about 3.7 million years old by Mary Leakey's team. Their discovery in 1978, which has delighted anthropologists, also causes problems because none of the hominins which have been discovered from this period could have made them.

About 3.6-7 million years ago Tanzania's Mount Sadiman erupted and showered the area with light volcanic ash. Subsequent rain allowed the ash to take imprints of the creatures which crossed it and about six or more distinct layers of about 6 inches thick were preserved.

Two footprint trails can be followed for about 50 meters (165 ft) and indicate two individuals about 1.4 meters (4'7") tall and the other 1.2 meters tall. In some places a third set of footprints is superimposed upon the larger ones, indicating to many that a family had once walked the area looking for food and water, or tried to flee the volcano.

The prints are unmistakably human and not those of an apelike hominin, with a big toe lying alongside the other toes and not splayed at an angle as in apes. The weight of distribution is the same as that of a modern unshod person, indicating a bipedal stride, rather than the rolling gate of an semi-arboreal ape.

Mary Leakey invited Russell H. Tuttle of the University of Chicago to study the footprints. In the March 1990 issue of 'Nature' magazine he concluded, "In sum, the 3.5 million year old footprints at the Laetoli site G resemble those of habitually unshod modern humans. None of the features suggest that the Laetoli hominids were less capable bipeds than we are. If the G footprints were not known to be so old, we would readily conclude that they were made by a member of our own genus, Homo. Thus we are left with something of a mystery as to the Laetoli hominids...The Laetoli footprints hint that at least one other hominid roamed Africa about the same time." ('Nature,' March 1990 p64)

One of the Laetoli footprints.

So, who walked across the volcanic ash

over 3.6 million years ago? Mary Leakey believed they were made by a non australopithecine ancestor of *Homo habilis*, whereas Johanson and White are sure they were made by *Australopithecus afarensis*, the same species as Lucy. Of course they all believed the creature had an apelike head with humanlike feet. Tuttle felt that *A. afarensis* could not have made the prints as the feet of this hominin and those of other australopithecines had long, curved toes.

The Laetoli footprints.

Stern and Susman objected and proposed that the ancient hominins had walked across the volcanic ash with their long toes curled under their feet, as chimpanzees were able to do. Tuttle found this explanation unlikely as there would be two patterns of toe impressions, long extended toes and short curled toes, with extra-deep knuckle marks. This was not the case with the Laetoli prints. Even Tim White, in favor of the *A. afarensis* model, saw no evidence of variation in lateral toe lengths to indicate knuckle walking.

The debate continues today but the current consensus is that australopithecines like Lucy made the prints, or perhaps the slightly more humanlike *Homo habilis*. This is based upon the age of the prints and the belief that Lucy was a biped.

Michael Cremo had a field day with the Laetoli prints and the academic controversy which followed them. He wrote, "Readers who have accompanied us this far in our intellectual journey will have little difficulty in recognizing the Laetoli footprints as potential evidence for the presence of anatomically modern human beings over 3.6 million years ago in Africa…What amazed us most was that scientists of worldwide reputation, the best in their profession, could look at these footprints, describe their humanlike features, and remain completely oblivious to the possibility that the creatures that made them might have been as humanlike as ourselves." (Cremo, M 'The Hidden History of the Human Race,' p262)

Was Australopithecus an ancestor? This is a hotly debated topic amongst anthropologists. The current consensus is that *A. afarensis* is the ancestor of both the robust australopithecines which later became extinct, and the Homo line. Others believe that *A. africanus*, evolved from *A. afarensis,* is a human ancestor. Louis Leakey felt that Australopithecus was an extinct ape and *Homo habilis* was the first true ancestor. Creationists do not believe that humans were descended from Australopithecus or any other ape.
REFERENCES: M. Cremo, 'The Hidden History of the Human Race'
S. Parker, 'The Dawn of Man'
Wikipedia articles on Australopithecus and Lucy
Wong, K., Footprints to fill, *Scientific American*, August 2005
Creation Ministries International, 'Was Lucy Bipedal?' Marvin Lubenow.
http://creation.com/was-lucy-bipedal-journal-of-creation-tj
'Nature Magazine,' March, 1990

HOMO HABILIS

Possibly the most hotly contested hominid species is *Homo habilis*, (handy man) which supposedly lived between 2.4 and 1.5 million years ago in Africa. Although similar to australopithecines in many ways, its face projected less than *A. africanus* and its back teeth were smaller but still larger than those of modern humans. With an average brain size at 650ccm, this species may also have been capable of rudimentary speech.

Original habilis specimens were discovered by Jonathon Leakey at Olduvai Gorge near the remains of *Zinjanthropus* in 1960. When Louis Leakey

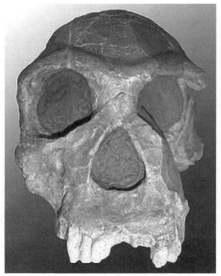

Homo habilis from Wikipedia.

discovered 'Zinj' near stone tools he said, "It is precisely by his manufacture of the first known pattern of implement that I believe *Zinjanthropus* can claim the title of earliest man at least until more distant toolmakers are found." However, the discovery of *Homo habilis* near the stone tools soon knocked Zinj off the throne of the oldest toolmaker.

Homo habilis reconstruction from Wikipedia.

The first hominid fragments (OH7) were lower leg bones, teeth, a finger, foot bone, braincase fragments and a mandible probably belonging to a juvenile of about 12 years old. The bones were thinner and lighter than Zinj, and the braincase fragments suggested a larger brain. As these more 'modern' fossils were found in older layers than Zinj, they were informally known as the 'pre-Zinj child.'

Eventually more fragments were discovered which were nicknamed Johnny's Child, George, Cindy and Twiggy. John Napier, an anatomist who studied the hand bones, agreed that the thumb was opposable so that the creature could have made the tools discovered at the site. Anatomist Michael Day declared that the foot bones resembled a humanlike foot, lacking the divergent big toe of apes and allowing bipedalism.

For many years anthropologists debated whether *Homo habilis* was actually *Australopithecus habilis*. This ignited another 'splitters vs lumpers' debate: the splitters tend to easily create new groups, species, subspecies and even families while lumpers try hard to fit new fossils into one of the existing categories. Lumpers wanted to place the new fossils into the australopithecine species or *Homo erectus* species which they believed evolved from it. Splitters argued successfully that fossils were sufficiently different from both the australopithecines and *Homo erectus* to merit a new species.

Louis Leakey had the reputation as a 'super splitter' who had created numerous new species (like Zinjanthropus) which had been incorporated into other groups. Palaeoanthropologist Le Gros Clark, on the other hand believed that *Homo habilis* was "easily incorporated within *Australopithecus africanus*" and was thus, not a separate species.

1470- *Homo habilis* or *Homo rudolfensis*?

One of the most important and controversial 'habiline' fossils was discovered in 1972 by Bernard Ngeneo of Richard Leakey's team at Koobi Fora on the east side of Lake Rudolf (now Turkana) in Kenya and became designated as KNM-ER 1470. It was discovered in volcanic tuff (known as the KBS tuff) which had been radio dated a few years previously at 2.6 million years old. But the dating of the KBS had many problems, as initially geophysicist Jack Miller had dated it at 200 million years old in a Cambridge lab! Further tests by Miller dated the KBS at 2.4 and 2.6 mya. As 1470 was uncovered in a lower level, it was presumed to be 3 million years old, an unacceptably ancient age.

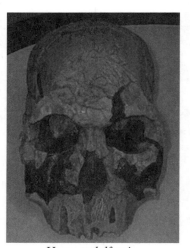

Homo rudolfensis.

Eventually 1470 was redated to 1.9 million because at three million years old it could not have been an ancestor, but was rather a contemporary of the australopithecines. The scientists assumed that the samples had been contaminated and had given a falsely ancient age.

1470 had some australopithecine characteristics such as a flattened face, but it also projected forwards to a degree. The upper jaw was relatively large and the teeth were comparable to modern humans. Its major feature was a high forehead and bulging, thin boned braincase, encasing a brain of about 800 ml.

The differences between 1470 and other habiline skulls eventually led to it being classified as a new species, *Homo rudolfensis* by V.P Alexeev in 1986. Although no associated postcranial remains have been discovered for 1470, it is assumed that it lacked features like slim hips for walking long distances and legs longer than arms.

In 2007 Dr. Timothy Bromage, a palaeonanthropologist at New York University College of Dentistry, studied *Homo rudolfensis* (1470) and claimed it was significantly more apelike than previously believed. His computer generated reconstruction shows the 1.9 million years old skull had a surprising small brain (about 530 ccm) and distinctly protruding jaw, features associated with earlier hominids. Bromage is the first to question Richard Leakey's depiction of *Homo habilis* as having a vertical facial profile and relatively large brain- an interpretation widely accepted until now. Mary Leakey had painstakingly constructed the cranium of 1470 from many fragments according to how she believed it had looked.

"Dr. Leakey produced a biased reconstruction based on erroneous preconceived expectations of early human appearance that violated principles of craniofacial development," said Dr. Bromage who presented his findings at the International Association for Dental Research in New Orleans in May 2007.

Dr. Bromage's reconstruction shows a sharply protruding jaw and a brain less than half the size of modern humans. This makes the skull far more like an australopithecine and less like a member of the Homo genus.

http://www.eurekalert.org/pub_releases/2007-03/nyu-med032307. php

Femurs and tali Some distance away from 1470 in the same stratum two human looking femurs were discovered by John Harris, a palaeontologist from the Kenya National Museum. These femurs, designated ER 1481 and ER 1472 were both attributed to *Homo habilis* despite the fact that Leakey claimed the leg bones were "almost indistinguishable from those of *Homo sapiens*." They were also more modern looking than *Homo erectus* bones, which seems highly anomalous. Furthermore, a talus (ankle bone) which was also discovered at Lake Turkana by B.A Wood shows a "similarity of KNM-ER 813 with modern human bones...not significantly different from the tali of modern bushmen." (Cremo, p253)

Michael Cremo, a skeptic of evolutionary theory wrote, "If the KNM-ER 813 talus really did belong to a creature very much like modern human beings, like the ER 1481 and ER 1472 femurs, into a continuum of such finds reaching back millions of years. This would eliminate *Australopithecus, Homo habilis and Homo erectus* as human ancestors." (M. Cremo, 'The Hidden History of the Human Race')

Current status

In 1987 the first *H. habilis* skeleton, OH 62, 'the Dik Dik specimen' was discovered at Olduvai Gorge with the bones of the body clearly associated with the skull. This apelike creature was only 3.5 feet tall and had long arms. Another find, a foot skeleton designated OH8 which was dated to 1.7 million years at Olduvai Gorge, resembled the foot of a chimp or gorilla. Furthermore, the anatomy of a hand called OH 7 found at Olduvai Gorge which some anthropologists considered humanlike, actually suggests that it was used to swing through trees.

H. habilis, hailed as the earliest toolmaker, has now been demoted by some to little more than an ape. ER 1470, once embraced as *H. habilis* has now been redesignated to *Homo rudolfensis* by Alexeev since 1985 because its face is unlike other habilines.

In his 1992 book 'Bones of Contention' creationist Marvin Lubenow argued that the habilis remains belonged to two separate species. 1470 belonged to a human while others like OH 24 were an ape species. He concluded that *Homo habilis* was not a valid species. This argument was echoed by Hindu creationist Michael Cremo who wrote: "Taking the many conflicting views into consideration, we find it most likely that the *Homo habilis* material belongs to more than one species, including a small, apelike, arboreal australopithecine (OH 62 and some of the Olduvai specimens), a primitive species of Homo (ER 1470 skull) and anatomically modern humans (ER 1481 and ER 1472 femurs.)" (ibid p 256) This of course begs the question, what were anatomically modern humans doing millions of years ago?

Another problem with the theory of linear evolution occurred when a *Homo habilis* skeleton, dated at 1.44 million years was younger than a *Homo erectus* found in Kenya dating to 1.55 million years in 2007. This find challenged the conventional view that these species evolved one after the other, whereas they apparently lived side by side in Eastern Africa for half a million years. It also suggests that both *habilis* and *erectus* must have originated from a common ancestor between two and three million years old.

According to Daniel Lieberman, a professor of biological anthropology at Harvard University, "Now we have extended the duration of the habilis species, and there's no doubt that it overlaps considerably with erectus."

The erectus skull was as small as a habiline, although its cranial ridge, jaw and teeth are all characteristics of erectus. This diminutive size makes it "not as humanlike as once thought." Dr. Lieberman commented, "The small skull has got to be female, and my guess is that all the previous erectus we have found turned out to be male."

Dr. Spoor, who along with Maeve and Louise Leakey reported on the find in 'Nature' journal, commented that the evidence clearly contradicted previous ideas of human evolution "As one strong single line from early to us."

BIBLIOGRAPHY

'Man's earliest direct ancestors looked more apelike than previously believed,' Press release, Christopher James, New York University, 24 March, 2007.
http://www.eurekalert.org/pub_releases/2007-03/nyu-med032307.php
M. Cremo, 'The Hidden History of the Human Race' op.cit

HOMO ERECTUS

According to evolutionary theory, Africa was the 'cradle of mankind' and eventually hominids spread out of Africa to the Middle East, Europe and Asia. However, it was in the far eastern Indonesian island of Java that palaeontologist Eugene Dubois decided to search for fossils of early humans because orang-utans and gibbons, distant relatives, could be found in the jungles.

INDONESIAN HOMO ERECTUS

JAVA MAN In 1892 Eugene Dubois discovered a fossilized human femur on the bank of the Solo River near Trinil in Java, Indonesia. The following year he discovered a skullcap and molar tooth from what he claimed to be the same individual about 45 feet away. Initially Dubois felt that he had discovered the cranium of an

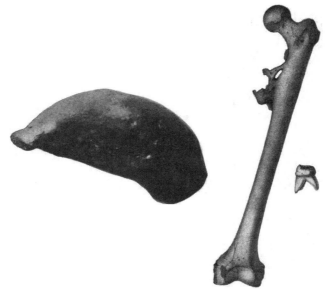

Bones of the Java Man.

ape, such as a large gibbon, but was persuaded by the famous Ernst Haeckel, Professor of Zoology at the University of Jena, that he had discovered 'the missing link', which he called *Pithecanthropus erectus*.

In 1894 he published a full report declaring, "*Pithecanthropus* is the transitional form which, in accordance with the doctrine of evolution, must have existed between man and the anthropoids." After returning to Europe in 1895 Dubois exhibited his bones in Paris, London and Berlin. When he presented his findings to the Berlin Society for Anthropology, Ethnology and Prehistory, its president Dr. Virchow declared the femur was human and the skull belonged to an ape. "The skull has a deep suture between the low vault and upper edge of the orbits. Such a suture is found only in apes, not in man. Thus the skull must belong to an ape. In my opinion this creature was an animal, a giant gibbon in fact. The thigh bone has not the slightest connection with the skull." (M. Cremo, 'The Hidden History of the Human Race,' p159)

An expedition by Professor Lenore Selenka to Trinil in 1907-8 recovered 43 boxes of fossils which contained evidence of human settlement such as charcoal and bones, but no new *Pithecanthropus* remains. Based upon this evidence, she concluded that humans and

Pithecanthropus were contemporaries.

For decades controversy around Dubois' claims continued with many eminent palaeontologists like Boule and Weidenreich claiming that the skullcap belonged to an ape. In 1932 Dubios and Dr. Bernsen discovered more femurs from a box containing specimens gathered from Trinil by Dubois' assistant Mr Kriele in 1900. Unfortunately they were not able to determine the position these femurs occupied in the original excavation, although later authorities assigned them to a particular stratum, despite the lack of in situ documentation.

Furthermore, dating of the original Java man is controversial. The geographical complexity of the island makes precise dating difficult. The Trinil fossils range in date from 1.8 million to maybe as young as 780,000 years old.

In his later years Dubois concluded that his skullcap belonged to a large gibbon but by that time *Pithecanthropus* was firmly entrenched in the ancestry of modern *Homo sapiens* and his proclamations were ignored. Furthermore, other *Homo erectus* femurs such as those found in China and Africa are substantially distinct from modern human femora. It is very possible that the Trinil femurs belong to anatomically modern human beings despite their date of 800,000 years old.

In 1931 Gustav von Koenigswald of the Geological Survey of the Netherlands East Indies was despatched to Java to search for more specimens after the discovery of Peking man in 1929. When a colleague found hominid fossils at Ngandong on the River Solo, Von Koenigswald classified them as Javanese Neanderthals. While Von Koenigswald and his Indonesian team did discover an upper jaw of an adult *H. erectus*, it was impossible to date it accurately.

Appointed to the Carnegie Institution in 1937, Von Koenigswald returned to Java with a generous grant. Hundreds of Javanese were hired and paid bonuses for discovering hominid bones. About 40 fragments which had

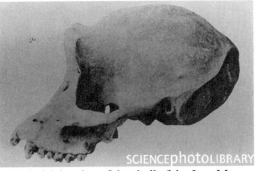

Dubois' drawing of the skull of the Java Man.

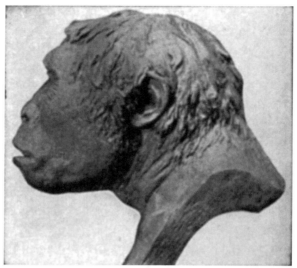

Early reconstruction of Java Man.

been pulled apart by the enterprising Javanese, were joined to form a *H. erectus* skullcap, he claimed, despite having no knowledge of its original stratigraphy. Dubois, who now believed his own skullcap belonged to an ape, accused von Koenigswald of fakery although he later believed his mistakes were probably not deliberate.

Von Koenigswald also discovered a gigantic lower jaw fragment which he called *Meganthropus palaeojavanics* (giant man of ancient Java) although he gave no description of the exact location at which it was found. Even larger fossilized human like teeth were attributed to a huge apelike creature called *Gigantopithecus*. While Von Koenigswald believed it was a large and relatively recent ape, Weidenreich proposed that both huge creatures were direct human ancestors. Most modern anthropologists believe that *Gigantopithecus* was an extinct giant ape while *Meganthropus* could be classified as either *H. erectus* or even *Australopithecus,* despite the fact that none of this species have been found out of Africa.

Other *H. erectus* classified remains have been discovered in Java. In 1963 T. Jacob reported a farmer finding a fossilized skull in the Sangiran area. Although he was unable to provide the exact position of the fragments when found, it was accepted as *Homo erectus* by anthropologists.

Other *Homo erectus* type fossils have been recently discovered in the Sambungmacan district of Java. According to a National

Geographic News article on February 27, 2003 by Hilary Mayell, scientists, led by Japanese anthropologist Hisao Baba analyzed the Sambungmacan 4 skull using digital visualization techniques. Baba and his colleagues argued that the morphological characteristics of the earlier Trinil fossils more closely resembled modern humans whereas later specimens from Ngandong (25,000 to 50,000 years old) were more primitive, or "an evolutionary tangent of its own, developing features that are not shared by modern humans." ('Java Skull Raises Questions on Human Family Tree')

CHINESE HOMO ERECTUS

YUANMOU MAN—HOMO ERECTUS YUANMOUNENSIS

The earliest dated hominin remains in China were discovered in the Yuanmou Basin, Yunnan Province in 1965 by a geological survey team when a railway was proposed from Chengdu to Kunming. Locals told the scientists where to find "dragon bones" which were in fact mammalian fossils, but had been used for countless years in Chinese medicine.

In one area the land was cut by a gully which left two walls standing exposed to erosion. The geologists could see fragments of bones in the two walls and one young geologist Qian Fang was startled to see human incisors embedded in the earth at the base of a four meter high mound.

According to Hu Chengzhi, a palaeontologist of the Museum of Geology of the Academy of Geological Sciences in Beijing, the two teeth are central upper incisors with well preserved crowns and highly fossilized. The teeth are big and strong and distinctly different from the incisors of *Homo sapiens*, indicating their extreme antiquity. Hu thought they were closer to the incisors of Peking Man at Zhoukoudian in their large size, robust build and complete pattern, but differed with tapering roots and a roughly triangular shape.

Palaeontologists Zhou Guoxing and Hu Chengzhi of the Chinese Academy of Sciences concluded that the incisors were more ancient than the fossils found at Zhoukoudian of 'Peking Man'. They wrote, "We have sufficient ground to assume that the central upper incisors discovered at Yuanmou are representatives of an earlier type of *Homo erectus* so far discovered in China. The morphological differences of

75

this genus from *P. pekinensis* are an indication of the primitiveness of the former, showing characteristics in the period of evolutionary transition from *Australopithecus africanus* to *Homo erectus*."

('Early Man in China,' Jia Lanpo, Foreign Languages Press, Beijing, 1980)

Lanpo said that the incisors were found in the 25th layer, fourth stratum, 600 meters from the base of the 695 meter thick 'Yuanmou Series' of deposits. Other mammalian fossils, including many now extinct, indicate that the incisors were found in deposits dating from between one to three million years old. Recent palaeomagnetic tests for the age of this series date the incisors at about 1.7 million years ago.

This date upsets the 'Out of Africa' hypothesis that *Homo erectus* evolved from *Homo habilis* in Africa about 1.5 million years ago. It also suggests the possibility that *Homo erectus* developed separately in China, independent of African influence, a very unpopular notion with evolutionists.

In 2004 a research team, including Richard Potts of the Smithsonian Institution's National Museum of Natural History, reported the results of excavating four layers in the Majuangou site in northern China.

The top layer, about 43 meters deep, contains the oldest record of hominin stone tools dating back to 1.32 million years ago, while the fourth and deepest layer in which the team found tools is about 340,000 years older than that.

According to Potts, "Because the oldest layers show humans made tools and extracted bone marrow like early people in Africa, the Majuangou evidence suggests strong connections with African hominins and their rapid spread across Asia."

All layers contain evidence that humans used stone tools such as choppers and scrapers. They also used their tools on bones of deer and horse shaped mammals, probably for food.

The research team used rock magnetic dating methods to establish the age of the artefacts collected at Majuangou and compared them to the soil history of a nearby site which had a more complete sediment record. Factoring other geological events such as magnetic pole movement, they pieced together a detailed age sequence of the archaeological levels.

No physical remains were uncovered although the tools were

assigned to *Homo erectus.*

'Science Daily' **http://www.sciencedaily.com/releases/2004/10/04 1001092127.htm**

PEKING MAN—HOMO ERECTUS PEKIENSIS

In 1921 Swedish geologist Johan Gunnar Andersson and American palaeontologist Walter Granger arrived in Zhoukoudian, near the capital of Peking to search for human fossils. Local quarrymen directed them to dragon Bone Hill where Andersson recognised deposits of quartz which indicated it was a rich area for fossils. The discovery of a fossilized human molar, followed by two more molars persuaded the Rockefeller Foundation to allot funding to the project.

The discovery of another tooth in 1927 by Davidson Black, of Peking Union Medical College, led him to identify a new species, *Sinanthropus pekinensis.* After his article was published in 'Nature', many fellow scientists were skeptical of Davidson's claims which were based on a single tooth, and the foundation threatened to withdraw funds.

Fortunately at the very end of the 1928 season, Pei Wenzhong made the historic find of an almost complete skull of *Sinanthropus* embedded partly in loose sands. Black was awarded an $80,000 grant that he used to establish the Cenozoic Research Laboratory.

According to Cremo; "With the financial backing of the Rockefeller Foundation for the Cenozoic Research Laboratory secure, Black resumed his travels for the purpose of promoting Beijing Man. He then returned to China, where work was proceeding slowly at Zhoukodian, with no new major *Sinanthropus* finds reported." (p197)

The discovery made Black (although not Pei) a media sensation and ensured continued access to the Rockefeller Foundation funds. Over the next few years Chinese archaeologists Yang Zhongjian, Pei Wenzhong

Reconstructed skull of Peking Man.

77

and Jia Lanpo uncovered 200 human fossils, including six almost complete skullcaps from more than forty individuals.

Reports showing extensive use of fire and the presence of stone and bone tools at Zhoukoudian were first published by Henri Breuil in 1931. This was unusual as evidence of fire usage had been omitted or overlooked in earlier reports. Black, Pei et al, embarrassed by this revelation, claimed that they had many doubts about the evidence of tool and fire usage so they did not report it. Or they could have deliberately withheld the information because it might have removed *Sinanthropus* from its unique position as China's earliest ancestor.

The Japanese occupation of 1937 put an end to further excavations at Zhoukoudian. Unfortunately the original specimens allegedly disappeared during the war en route to the port city of Qinghuangdao although casts remain.

The discovery of Peking man was fortunate for Java Man, who was upgraded to the genus Homo based upon its supposed fire making and hunting abilities. Peking and Java men were both classified as *Homo erectus*, fire and tool maker.

However, its status of fire maker has been challenged on various occasions. Binford and Ho, anthropologists at the University of New Mexico, claimed the ash deposits were actually huge guano droppings inside the cave which could have burned. "The assumption that man introduced and distributed fire is unwarranted, as is the assumption that burned bones and other materials are there by virtue of man's cooking his meals." (Cremo, p199) In 1998 Steve Weiner of the Weizmann Institute of Science also came to a similar conclusion.

AFRICAN HOMO ERECTUS

The first fossils of *Homo erectus* in Africa were discovered in the 1950s by French palaeontologist Camille Arambourg. From 1954-5 three adult mandibles were found at Ternifine, near Oran in Algeria. A partial skull was also found with milk teeth from a child. These mandibles had the characteristic chinless configuration of Asian *Homo erectus*.

HOMO ERGASTER
This hominin lived in eastern and Southern Africa between 1.9

and 1.4 million years ago. The species was named by Groves and Mazak in 1975 after the Greek word 'ergaster' meaning 'Workman'. Ergaster was chosen due to the discovery of various 'advanced' tools such as hand-axes and cleaves near the skeletal remains of a fossil. These tools have been classified as 'Acheulean', the same used by *Homo erectus.*

Homo ergaster.

Homo ergaster is distinguished from *Homo erectus* by its thinner skull bones and lack of a sulcus. It also has a smaller, more othognathic face, smaller teeth and a larger (700 and 850ccm) cranial capacity. However it stood very tall at 1.9 m (6ft 3ins.)

The most famous and complete skeleton was discovered at Lake Turkana, Kenya in 1984 by palaeoanthropologists Richard Leakey, Kamoya Kimeu and Tim White. This 1.6 million year old specimen was named KNM-WT 15000, and nicknamed 'Turkana Boy.'

There is considerable controversy whether *Homo ergaster* is a separate species or a variation of *Homo erectus*. Another fossil KNM-ER 3733 has been variously labelled as either *H. ergaster* or *H. erectus* and is dated at 1.75 mya. This hominin is a contemporary of *Homo habilis* and various robust australopithecines, which shattered the theory that only one species was alive at any one time. Furthermore, some researchers see ergaster as a direct ancestor of modern humans with erectus being an evolutionary dead-end, despite the fact that most ergaster hominins are older than less advanced erectus.

Homo erectus fossils have also been found at Olduvai Gorge by Louis and Mary Leakey, including a 1 million year old skull. The Swartkrans caves in South Africa also contain *Homo erectus* fossils in later upper levels.

HOMO ERECTUS IN EUROPE

Although many tools of the Acheulean designation have been found in Europe, fossilized remains of *Homo erectus* are much rarer.

The most famous would be the robust mandible found in Mauer, Germany which has been called 'Heidelberg Man.' This mandible indicates that a very robust creature lived in that area about 400,000 years ago.

HEIDELBERG MAN (*Homo heidelbergensis*)

In 1907 Daniel Hartmann, a sand pit workman at Mauer, Germany, discovered a massive hominid jawbone at a depth of 82 feet. Its premolar teeth were missing but later recovered. Dr. Otto Schoetensack of the University of Heidelberg's Geology Department designated it *Homo heidelbergensis* and dated it between 250,000 and 450,000 years old.

The Heidelberg mandible is still controversial as its thickness is common to *H. erectus*, while its teeth and cusp patterns are very modern in size. It is thought to be a direct ancestor of modern *Homo sapiens*, with a tool technology similar to the Acheulean tools of *Homo erectus*.

Other designated remains have been discovered at Arago, France and Petralona, Greece. Rhodesian Man, discovered in Africa, is also believed to be within the *Homo heidelbergensis* species. Both *H. heidelbergensis* and *H. antecessor* are believed to be descended from the similar *Homo ergaster* from Africa, although the European species had a larger brain case, from 1100-1400 cm3 which is within the parameters of modern humans.

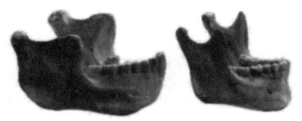

Heidelberg Man mandibles vs a modern human.

HOMO GEORGICUS

By far the most controversial and oldest find in Europe was in the Caucasian country of Georgia in 2001 where three hominid skulls were discovered and dated to about 1.8 million years old, pushing back the so-called human exodus from Africa by about one million years. Two skulls, which are very well preserved, have been classified as *H. erectus*, whereas the third skull has more primitive characteristics of a *Homo habilis* which has never been found out of Africa. Thousands of primitive chopper like tools, traditionally associated with *H. habilis,* were found in the vicinity.

David Lordkipanidze, deputy director of the Georgian State Museum at Tblisi, led the international team which had been excavating at the border of South-Caucasus since 1991. On the last day of excavations a hominid mandible, known as D211, was discovered.

Initially the hominid remains were dated to 1.6 million years and were assigned as *Homo erectus.* But in 2001 they were redated to 1.810,000 years.

The third skull of a similar age is much smaller, at 600 ccm, while the other skulls are 650 ccm and 780 ccm. "These hominids are more primitive than we thought, something intermediate – between *H. habilis* and *H. erectus*. It could even be one of the earliest Homos in the world," said Dr. Lordkipanidze to National Geographic.

This third skull has inspired excited comments from palaeontologists such as Dr. Ian Tattersall from the Natural History Museum in New York City who called it, "the first truly African-looking thing to come from outside Africa."

Dr. Fred Smith of Loyola University urged caution, doubting that two separate hominid species could have occupied the same habitat at roughly the same time.

"The possibility of variations within a species should never be excluded," Dr. Smith said. "There's a tendency now for everybody to see three bumps on a fossil instead of two and immediately declare that to be another species."

Some discoverers of the Dmanisi skull speculated that these hominids are direct descendants of *Homo habilis* which had already left Africa. Indeed, the tools uncovered are similar to those found at Oldowan in Africa which have been assigned to *Homo habilis*. In that case, *Homo erectu*s may not have evolved in Africa but elsewhere which would alter the human family tree. Other researchers like Lordkipanidze believe it is a new species, '*Homo Georgicus'*.

By 2009 five individuals had been excavated, and at the age of 1.8 million years, are the oldest hominins discovered outside Africa.

Homo georgicus.

This discovery directly challenges the ROA theory that our ancestors evolved in Africa and then spread to different continents. In fact it suggests that some of our oldest ancestors lived in the Caucasus nearly two million years ago.

Anthropologists, however, have to tailor the evidence to fit their preconceived theories. According to Lordkipanidze, some *Homo erectus* may have left Africa for Eurasia before returning much later. That would neatly dovetail the ROA theory with the Dmanisi fossil evidence.

Chris Stringer, head of human origins at the Natural Museum in London commented, "The Dmanisi fossils are extremely important in showing us a very primitive stage in the evolution of Homo erectus. They raise important questions about where that species originated."

http://www.guardian.co.uk/science/2009/sep/08/fossils-georgia-dmanisi-early-humans

'Fossil find in Georgia challenges theories on early humans,' Ian Sample, guardian.co.uk September 8, 2009.

The Dmanisi people stood about 1.5 meters tall and had brains about half the size of those in modern humans. According to Lordkipanidze, they were "almost modern in body proportions and were highly efficient walkers and runners. Their arms moved in a different way, and their brains were tiny compared to ours. They were sophisticated tool makers with high social and cognitive skills." (ibid)

HOMO ANTECESSOR is classified as the second oldest hominin in Europe dating from 1.2 million to 800,000 years old. It was discovered by Eudald Carbonell, J. Arsuage and M. Bermudez de Castro in Spain and named by Castro in 1997.

The best preserved fossil is a maxilla found in Spain dating from 780,000 to 857,000 years old. *H. antecessor* stood from about 1.6 to 1.8 meters tall, weighed about 90kg and had brain sizes from 1000 to 1150 ccm. It also had a protruding occipital bun, a low forehead and lack of chin.

Over 80 bone fragments from six individuals were discovered at the Gran Dolina site in 1994 and 1995. Furthermore, roughly 200 stone tools and about 300 animal bones were discovered at this site which is dated to about 780,000 years old.

It is interesting to note that some of the remains are almost indistinguishable from the KNM-WT 15000 Turkana Boy which is attributed to *Homo ergaster*, a species that has not been discovered outside Africa.

There are controversies surrounding the classification of *H. antecessor*. Some anthropologists believe that *H. antecessor* is either the same species or a direct antecedent to *Homo heidelbergensis* (Heidelberg Man) or Neanderthal man. Bermudez de Castro and his colleagues believe that a few dental and cranial features suggest *Homo antecessor* is closer to *Homo ergaster*. They argue that while *Homo antecessor* has similarities to later *Homo heidelbergensis*, it has more in common with modern humans than does Heidelberg man. According to their theory, *Homo ergaster* gave rise to *Homo antecessor* in Africa which about one million years ago spread to the Middle East and Europe, including Gran Dolina. In Europe it evolved into *Homo heidelbergensis* and then to Neanderthals, whereas in Africa it evolved into Homo sapiens, by passing both *Homo erectus* and *Homo heidelbergensis* which are not part of the human lineage.

However, many palaeoanthropologists, such as Philip Rightmire of the State University of New York, Christopher Stringer of the Natural History museum in London and Milford Wolpoff of the University of Michigan, have expressed reservations about the designation of *Homo antecessor* and the revision of the evolutionary tree which ejects *Homo erectus*. They claim that the "modern" features are in fact juvenile traits not present in adults, although Antonia Rosas of the National Museum of Natural Sciences in Madrid counters that the Gran Dolina fragments are adult and do show modern traits.

While welcoming the controversy, Arsuaga has commented, "Our opinion is that *Homo heidelbergensis* is a Middle Pleistocene European species which is ancestral only to Neanderthals and not to modern humans. The last common ancestor to Neanderthals and modern humans is older, and is the newly named species *Homo antecessor* (of Lower Pleistocene age)."

'Archaeology,' A New Species? **http://www.archaeology.org/ online/news/gran.dolina.html**

HOMO ERECTUS TIMELINE

According to Wikipedia, the current timeline for *Homo erectus*, also known as *Pithecanthropus,* is that it existed from 1.8 million to 300,000 thousand years ago in Africa, Europe and Asia. The face has protruding jaws with large molars, thick brow ridges, no chin and a brain size varying between 750 ccm and 1225 ccm. Erectus probably used fire and its stone tools, known as the Acheulean assembly, were more sophisticated than those of habilis.

- *Homo georgicus* was discovered in 2001 in Georgia and is about 1.8 million years old. The skull with a brain capacity of 600ccm suggests that this fossil could be a blending of *H. habilis* and ergaster races.
- Turkana boy is sometimes classified as the subspecies *Homo ergaster*. Discovered in Kenya in 1984, it is an almost complete skeleton of a twelve year old boy, minus hands and feet. The boy, 160 cm tall (5'3") would have been about 185 cm (6'1") as an adult, much taller than earlier homo specimens. Turkana boy lived about 1.6 million years ago.
- Sangiran 17 is the most complete erectus fossil from Java. It has been dated from 800,000 to 1.7 million years old.
- In 1960 Louis Leakey discovered a skeleton known as 'Chellean Man' at Olduvai Gorge. It is almost 1.5 million years old and had massive brow ridges with a brain size of 1065 ccm.
- Heidelberg Man was discovered in 1907 in Germany and is between 400,000 and 700,000 years old. With a lower jaw and all its teeth, this specimen seems to be at the small end of the erectus range.
- Java man is about 700,000 years old. The skeletal remains consisting of a thick skull cap, teeth and femur may have been mixed up with modern remains.
- Peking man- Between 1929 and 37 fourteen crania, 11 lower

jaws, teeth, bones and tools were discovered at Zhoukoudian (Choukoutien) near Beijing (Peking). They are estimated to be from 500,000 to 300,000 years old. Unfortunately the bones disappeared in 1941 although casts and descriptions remain.

- Rhodesia Man was discovered in 1921. With its complete cranium with large brow ridges and receding forehead, its age is between 200,000 years and 125,000 years. The brain size was 1280 ccm.
- Petralona 1, discovered in Greece in 1960, shows features of being part erectus and part Neanderthal with a brain size of 1220 cc. It is from 250,000 to 500,000 years old.

SOME DATING ISSUES

In recent years, debate has been raging about the age of *H. erectus* fossils recovered from some sites in Java. In 1996 Susan Anton from New York University reported to the annual meeting of the American Association of Physical Anthropologists that fossil bearing sediment containing *Homo erectus* remains had been dated to between 50,000 and 30,000 years ago. These radiocarbon readings suggested that *Homo erectus* had survived well into the era dominated by modern humans.

A new analysis, based upon measurements of radioactive argon's decay in volcanic rock surrounding the fossils gave the date of 550,000 years. Anton was not sure why the estimates differed so dramatically and which one is more accurate. Christopher Stringer commented that animal fossils on Java date from 200,000 to 150,000 years ago and provide a possible framework for when *H. erectus* lived there.

If nothing else, it highlights the fact that dating can be an inexact science.

REFERENCES:
National Geographic, February 27, 2003 by Hilary Mayell 'Java Skull Raises Questions on Human Family Tree'
M. Cremo, 'The Hidden History of the Human Race'
Jia Lanpo, 'Early Man in China' Foreign Languages Press, Beijing, 1980)
'Fossil find in Georgia challenges theories on early humans,' Ian

Sample, guardian.co.uk September 8, 2009.
http://www.guardian.co.uk/science/2009/sep/08/fossils-georgia-dmanisi-early-humans
'Archaeology', A New Species? Mark Rose, July 29, 1997
http://www.archaeology.org/online/news/gran.dolina.html
Wikipedia entry on Homo erectus
Science News, Oct 1, 2004 'Out of Africa: Scientists Find Earliest Evidence Yet of Human Presence in Northeast Asia,' **http://www.sciencedaily.com/releases/2004/10/041001092127.htm**
Science News, April 16, 2010, 'Java Man takes age to extremes,' Bruce Bower
http://www.sciencenews.org/view/generic/id/58346/title/Java_Man_takes_age_to_extremes

ARCHAIC HOMO SAPIENS

The latest *Homo erectus* fossils have been dated to around 200,000 years ago so that lineage either evolved into *Homo sapiens* or became extinct. Most creationists believe that the *Homo erectus* lineage became extinct, but they ignore the numerous fossilized remains of transitional hominins in Europe, Asia and Africa.

Transitional features include some of those of *Homo erectus* such as large upper jaws and jutting brows with *Homo sapiens* features such as a larger brain and prominent chin. The most famous archaic *Homo sapiens* is the Neanderthal species.

ARCHAICS IN AFRICA

'Broken Hill man' (also known as *Homo rhodesiensis*, Kabwe skull) from northern Rhodesia (now Zambia) was the first archaic fossil to be recovered in Africa. Uncovered from a mine in 1921, this hominin had thicker heavier facial bones and prominent brow ridges with a brain size of about 1,100 ccm and badly decayed small teeth.

At least two skulls as well as two hip bones and arm and leg fragments were also recovered from the zinc mine. The length of the tibia put the body height at about 1.7 meters.

Initially Broken Hill man was dated from 300,000 to 100,000 years old but it has been redated to somewhere between 150,000-

130,000 years old. However absolute stratigraphic dating is impossible because the mine is now filled with water.

This hominin presents some problems for anthropologists. It was an extremely robust individual with comparatively the largest brow-ridges of any hominin remains. With its broad face it has been described as an African Neanderthal, but the current consensus is that Rhodesian man is within the group of *Homo heidelbergensis,* (Heidelberg man) another robust species. According to Tim White, it is the ancestor of *Homo sapiens idaltu* which itself is supposedly the ancestor of *Homo sapiens sapiens.*

Homo rhodesiensis.

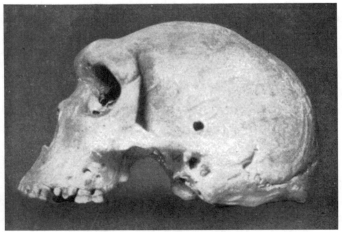

Side view of *Homo rhodesiensis.*

Other archaic Homo sapiens have been found at:
- Ndutu, Tanzania- heavy brow ridges with a rounded cranium.
- Bodo, Ethiopia- an extremely robust individual aged from 400,000-300,000 years.
- Elandsfontein, South Africa- less marked eyebrow ridges, probably a female.
- Laetoli- domed head- mixed features about 120,000 years old.

87

- Omo1, Ethiopia- domed skull and well formed chin about 130,000 years old.
- Omo 11- Erectus features, sagittal crest, apparently a contemporary of Omo 1.

HOMO SAPIENS IDALTU is officially known as the oldest 'anatomically modern' human and consists of fossilized skulls of two adults and one child. It was discovered in Ethiopia's Afar triangle by Tim White in 1997 and unveiled in 2003. Radioisotope dating indicates the three crania are between 154,000 and 160,000 years old. One adult male (BOU-VP-16/1) had a cranial capacity of 1450 ccm but possessed 'archaic features,'

These fossils date precisely from the time that biologists predicted that a genetic 'Eve' lived in Africa who gave rise to all modern humans. According to White, "We've lacked intermediate fossils between pre-humans and modern humans, between 100,000 years ago and 300,000 years ago and that's where the Herto fossils fit....Now the fossil record meshes with the molecular evidence."

'160,000-year-old fossilized skulls uncovered in Ethiopia are oldest anatomically modern humans,' 'UCBerkley News' Robert Sanders, 11 June 2003

http://www.berkeley.edu/news/media/releases/2003/06/11_idaltu.shtml

These hominins lived long before most examples of Neanderthals, proving beyond a doubt that *Homo sapiens* were not descended from their stocky cousins. Clark Howell, another member of the research team commented, "These fossils show that near-humans had evolved in Africa long before the European Neanderthals disappeared. They thereby demonstrate conclusively that there never was a Neanderthal stage in human evolution." (ibid)

Idaltu means 'elder' in the Afar language and refers to the adult male who was about 30 but had heavily worn upper

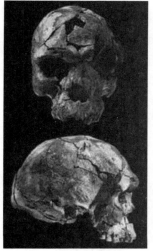

Homo sapiens idaltu.

teeth and a brain slightly larger than our modern brain.

ARCHAICS IN CHINA

China continually provides palaeontologists with fossils of great interest. Sometimes these early hominin fossils seem to challenge the 'Out of Africa' hypothesis and favor the 'Multiregional Continuity Model,' or even resurrect the old 'Out of Asia' theory.

In 1956 a peasant found a transitional skull near Maba, Guangdong province. Due to its modern characteristics, it was determined to be a *Homo sapiens* from the Late Pleistocene. According to Cremo, it was only dated from the Late Pleistocene because of the "morphology of the hominid remains," and could have been from an earlier period based upon the animal bones found in close proximity.

Another example of dating by morphology would be the discovery of two *Homo sapiens* teeth found near Tongzi, Guizhou province in the early 1970s. Even though the fauna suggests a Middle Pleistocene range, the human evidence, in accordance with evolutionary theory, allowed the teeth to be dated to a later period. Cremo, on the other hand, studied the reports of faunal remains from the site and determined that the Tongzi teeth belonged in the Middle Pleistocene, or contemporaneous with the Zhoukoudian *Homo erectus*.

He wrote, "We are not insisting that these beings actually coexisted. Perhaps they did, perhaps they did not. What we are insisting on is this- scientists should not propose that the hominids definitely did not coexist simply on the basis of their morphological diversity. Yet this is exactly what has happened. Scientists have arranged Chinese fossil hominids in a temporal evolutionary sequence primarily by their physical type. This methodology insures that no fossil evidence shall ever fall outside the realm of evolutionary expectations." (Cremo, p208)

In 1957 Chinese archaeologists under the leadership of palaeoanthropologist Jia Lanpo found fossils of Changyan man. The maxilla, judged *Homo sapiens* with some primitive features, was found in association with Middle Pleistocene fauna. Changyan Man was dated to about 200,000 years old because of these morphological characteristics, not because of the stratigraphy. The same classification occurred with fossils found at Liujiang cave in Guangxi Autonomous Region in 1958. The skull, vertebrae, ribs, pelvic bones and femur were anatomically modern but found in Middle Pleistocene deposits with fauna of that era. Chinese scientists, just like their Western contemporaries, assigned it to the Late Pleistocene because of its morphology.

Other early *homo sapiens* skull in China include 'Hsuchiayao Man,' a group of fragments uncovered near Datong in Shanxi and estimated to be about 100,000 years old.

Another more complete skull was unearthed at Shaoguan in Guandong which had a rounded cranium with pronounced brow ridges and thick bone typical of erectus. In 1978 another site in Shanxi, at Dali, yielded a fossil skull with intact facial bones which showed a combination of old and new characteristics.

In 2005 Chinese archaeologists announced evidence of early *Homo sapiens* in the valley of Qingjian River, challenging the 'Out of Africa' hypothesis that modern humans originated in Africa about 100,000 years ago.

Chinese palaeontologists are very interested in the fossil hominin records of 100,000 years ago because it is supposed to be a critical time for the expansion of modern humans from Africa, according to the currently favored 'Out of Africa' hypothesis.

In 2007 researchers of the Institute of Vertebrate Paleontology and Paleoanthropology, Chinese Academy of Sciences and Chongzuo Biodiversity Research Center, Peking University, excavated a 110,000 year old hominin site at Mulan Mountains in Chongzuo, Guangxi. Two hominin teeth and a mandible were unearthed as well as many other mammalian fossils such as *Gigantopithecus*. The fauna associated dates it to the late Pleistocene. Palaeoanthropologist Xinzhi Wu decided that the morphological features of the mandible placed it as an early *Homo sapiens*. These features include a protruding chin like that of *Homo sapiens* but a thickness of the jaw which is indicative of earlier hominins.

On Nov 3, 2009, 'New Scientist' ran an online article called 'Chinese challenge to the 'Out of Africa' theory stating that the find could lend support to Wolpoff's 'multiregional hypothesis.' Wolpoff wrote that the paper "acts to reject the theory that modern humans are of uniquely African origin and supports the notion that emerging

African populations mixed with natives they encountered."

Other anthropologists like Erik Trinkaus of Washington University in St Louis, Missouri questioned whether it is a *Homo sapiens* or a more archaic species. Chris Stringer of London's Natural History Museum concurred. "The fossil could just as likely be related to preceding archaic humans, or even to the Neanderthals, who at times seem to have extended their range towards China."

http://www.newscientist.com/article/dn18093-chinese-challenge-to-out-of-africa-theory.html

Xinzhi Wu and other Chinese scientists propose a 'Continuity with hybridization' model for the origin of modern humans in China, believing that the "the chronologically continuous hominin fossil records for China supports the argument that the modern humans of China are descendants of early hominins, like Peking Man, who had colonized the area, but already carried some foreign genes."

'The Homo sapiens cave hominin site of Mulan Mountain, Jiangzhou District, Chongzuo, Guangxi with emphasis on its age', Jin ChangZhu et. al.

http://www.springerlink.com/content/26m138v171861478/fulltext.pdf

Like the bigger war between evolution and creationism, the battle between the 'Out of Africa' hypothesis and 'Multiregional Continuity Model' is ongoing, especially in China.

ARCHAICS IN EUROPE

The Petralona skull, uncovered from a limestone cave in Greece in 1960, showed a mixture of features from *Homo erectus* and *Homo sapiens*. It has a flat, thick, erectus like cranium but in other respects it has *Homo sapiens* features and is dated to about 300,000 years.

The Swanscombe skull was unearthed from gravel pits near the British town of Swanscombe in 1933. Parts of the skull which were uncovered were the occipital bone, skull case and the left parietal bone. Reconstruction of the whole braincase gave a brain volume of about 1,325 millilitres.

Swanscombe skull.

91

Unfortunately the front of the skull, including the brow and chin were missing. This person lived about 300,000 years ago and remains the earliest Briton ever discovered.

The fossilized Steinheim skull was uncovered at gravel pits near Stuttgart, Germany and is estimated to be from 350,000 to 250,000 years old. The high domed skull is slightly flattened and has a cranial capacity of 1110 to 1,200 ccm.

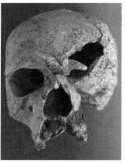

Steinhem skull.

REFERENCES:
'160,000-year-old fossilized skulls uncovered in Ethiopia are oldest anatomically modern humans,' 'UCBerkley News' Robert Sanders, 11 June 2003
http://www.berkeley.edu/news/media/releases/2003/06/11_idaltu.shtml
M. Cremo, 'The Hidden History of the Human Race'
Phil McKenna, 'Chinese challenge to the 'out of Africa' New Scientist Nov 3, 2009
http://www.newscientist.com/article/dn18093-chinese-challenge-to-out-of-africa-theory.html
'The Homo sapiens cave hominin site of Mulan Mountain, Jiangzhou District, Chongzuo, Guangxi with emphasis on its age,' Jin ChangZhu et. al.
http://www.springerlink.com/content/26m138v171861478/fulltext.pdf

NEANDERTHAL MAN

Neanderthals are either classified as a subspecies of humans (*Homo sapiens neanderthalensis*) or as a separate species (*Homo neanderthalensis*). They appear to have been indigenous to Europe and the Middle East for a span of about 100,000 years from 130,000 to 30,000 years ago.

The first Neanderthal remains were discovered in the Neander Valley in Germany in 1856, but prior to this date skulls had been discovered in Engis Belgium and Forbes' Quarry, Gibraltar.

Drawing of a Neanderthal.

Neanderthals were a few inches shorter than humans- the males were about 164 to 168 cm (5'4" to 5'6") and the females from 152 to 156 cm (5' to 5'1.5"). Their build was robust and they were undoubtedly muscular. Their skulls were low and flat, with an occipital bone or 'bun' at the back. The brow ridges were prominent and the chin receding.

Physically they had long collar bones, wide shoulders, a barrel-shaped rib cage, short shoulder blades, large kneecaps, short shinbones and possible bowed femurs. Recent research indicates that some Neanderthals had white skin and red hair.

The Neanderthal brain was surprisingly large- larger in fact than that of many people alive today. Its size ranged from 1350 to 1700 ccm, with 1450 being the average. Female Neanderthal brains were about 200 ccm smaller than those of males as their bodies were also smaller. Modern female brains are also about 10% smaller than those of males because they are generally smaller in build.

The following list from Wikipedia shows a timeline of Neanderthal fossil discoveries.

- 1829- Skulls found in Engis, Belgium.
- 1848- Skull from Forbes' Quarry, Gibraltar.
- 1856- Johann Karl Fuhlrott recognized 'Neanderthal man' fossil found in the Neandertal Valley near Mettmann.
- 1866 - Two nearly perfect male and female skeletons found at Spy, Belguim.
- 1908- A nearly complete skeleton was discovered with Mousterian tools and the bones of extinct animals discovered at Le Moustier, France.

- 1908- Fossilized skull found at La Chapelle-aux-Saints, France 60,000 years old. Specimen was toothless and arthritic.
- Ralph Solecki uncovered 9 Neanderthal skeletons in Shanidar cave, Iraq.
- 1975- Erik Trinkaus' study of Neanderthal feet confirmed they walked like modern humans.
- 1987- Israeli fossils at Kebara were dated to 60,000 B.P. Qafzeh to 90,000 B.P. and Es Skhul 80,000 B.P.
- 1991- ESR dates show the Tabun Neanderthal was contemporaneous with modern humans from Skhul and Qafzeh.
- 1997- Neanderthal mitochondrial DNA (mtDNA) uncovered from Neander Valley fossil.
- 2000- DNA uncovered from Late Neanderthal infant in Caucasus.
- 2005- Max Planck Institute for Evolutionary Anthropology launched project to reconstruct the Neanderthal genome.
- 2009 'First draft' of complete Neanderthal genome completed.

NEANDERTHAL CONTROVERSIES

Where they human ancestors? The discovery of *Homo sapiens idaltu* in Ethiopia in 1997 knocked the Neanderthals out of the human evolutionary tree as they had lived before them and possessed more modern physical characteristics.

Furthermore, several important sites such as Qafzeh Cave, Israel, suggest that Neanderthals arrived in the area after modern *Homo sapiens*. This indicates while there was some coexistence, there was probably little interbreeding between the two.

HOW DID THEY DIE OUT?

Until recently it was assumed that Neanderthals evolved into modern humans and did not become extinct. By 40,000 years ago *Homo sapiens* and Neanderthals had shared similar habitats for tens of thousands of years, but soon after Neanderthals disappeared from the archaeological record.

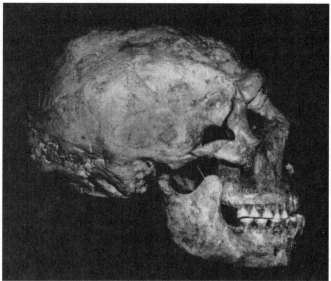
Neanderthal skull from Shanidar Cave, Iraq.

A cave at Ortvale Klde in Georgia was occupied by both Neanderthals and *Homo sapiens* between 50,000 and 21,000 years. While they both hunted the same animals, the tools of the *Homo sapiens* were more sophisticated, leading to the hypothesis that *Homo sapiens* survived because they were superior hunters.

Another theory proposes that Neanderthals bred with *Homo sapiens* and disappeared through absorption about 35,000 years ago. This was favored by Neanderthal expert Dr. Eric Trinkaus who has studied a child's skeleton from the Abrigo do Lagar Velho in Portugal which, at 25,000 years old, shows both *Homo sapiens* and Neanderthal features.

However, according to Dr. Briggs of the Max Planck Institute, "What we've done is confirm that the mitochondrial DNA of Neanderthals and modern humans was so different that it forms powerful evidence that there was very little if any interbreeding between the two species," said Dr. Briggs.

"We have also got tantalizing evidence that the Neanderthals formed a small population and we can only speculate as to what happened to them. Small population sizes are always more prone to extinction and they have a greater chance of something going wrong."

Current research, such as the Neanderthal Genome Project, favors the hypothesis that Neanderthals and humans were not related but shared a common ancestor about 500,000 years ago and that the Neanderthals did not contribute to the modern gene pool.

Recreation of a Neanderthal child.

WHAT DOES THE NEANDERTHAL GENOME PROJECT REVEAL?

In 2006 the Max Planck Institute for Evolutionary Anthropology and 454 Life Sciences announced they would sequence the Neanderthal genome which consists of about three-million base pairs. Only one Neanderthal specimen had enough DNA to sample— a 38,000 year old femur from Vindija cave, Croatia although three more specimens were later used. Preliminary studies by director Professor Svante Paabo indicated that this Neanderthal shares about 99.5% of its DNA with modern humans, but this is not enough to indicate it is of the same species. MtDNA analysis suggests that Neanderthals and modern humans shared a common ancestor about 500,000 years ago.

Edwin Rubin of the Lawrence Berkeley National Laboratory in Berkeley, California stated that recent genome testing suggests that human and Neanderthal DNA are from 99.5 to 99.9% identical.

In 2009 Richard Green et al. from the Max Planck Institute published the first draft of the full sequence of Neanderthal mitochondrial DNA. It was announced by that the Neanderthals had lived in small and isolated populations and probably did not interbreed with their human neighbors. The gene sequence for lactose intolerance was also discovered, showing that the Vindjina Neanderthal was unable to digest milk in adulthood.

RECENTLY DISCOVERED FOSSILS OF EARLY HOMININS

'TOUMAI MAN' (Sahelanthropus tchadensis)

In July 2002 the British scientific journal 'Nature' announced that the oldest member of the human family had been discovered in Chad, Africa by Professor Brunet of the University of Poitiers. The skull and jaw fragments of the earliest hominid ever discovered were announced as being six to seven million years old. This fills in the 'missing link', the vital five million year gap in the human family tree between accepted ancient apes and accepted ancient hominids.

Brunet claimed that Toumai had small canines, large molars and premolars that had thick enamel, similar to later hominids. The 'foramen magnum' the opening at the base of the skull where the spinal cord enters, suggested that Toumai walked upright.

Other scientists are not convinced that Toumai is an early hominid but rather an ape. Brigitte Senut of France's National Museum of Natural History, Martin Pickford of the College of France and John Hawks of the University of Wisconsin are among the detractors. They believe Toumai's teeth are more apelike than human like, whereas its massive brows are not seen in the earliest known hominids.

Dr. Wolpoff, of the University of Michigan believes it is a fossilized ape which was not a biped, a view shared by Vedic creationist author Michael Cremo.

Brunet responded, "Wolfpoff et al have described no derived ape feature of Toumai, nor have they disproved any derived features that this species shares with later hominids."

Henry Gee, senior editor of Nature, said he had visited Poitiers twice to see the skull and had Brunet's paper anonymously peer reviewed by five independent scientists. "Whatever Toumai is, it shows a combination of features we haven't seen before in hominids or apes."

http://www.whyevolution.com/toumai. html

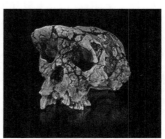

Toumai Man.

97

Author Michael Cremo has written a book on Anomalous Archaeological findings called 'Forbidden Archaeology.' He believes that Toumai man is simply another variety of ape which coexisted with humans millions of years ago. He says, "It's another example of the process of 'knowledge filtration' upholding the ruling paradigm. Every few months some scientist announces the discovery of fragmentary bones of some apelike creature with an age of a few million years. These are portrayed as earthshaking finds destined to revolutionize our entire picture of human origins. Actually this Toumai find is very apelike."

In 2008 a fresh controversy about the Toumai fossil erupted when radiological measurements estimated that the soil where it was found was between 6.8 and 7.2 million years old. The skull's discoverer, Alain Beauvilan, of the University of Paris at Nanterre, has now publicly challenged this estimate. Declining to participate in the hominid vs chimp debate, he says that contrary to Brunet's assertions that the fossil had been 'unearthed', it was found loose in the sand. A thick blue iron based mineral encrusted the skull, which also showed signs of weathering from the desert.

"How many times was it exposed and reburied by shifting sands before being picked up?" he asked in a commentary in the South African Journal of Science.

Furthermore, he says that the soil samples were taken selectively and did not give a full picture of the depth and range of the original topography. Some of the collection choices were described as "astonishing."

Beauvilan attacks Brunet's dating of another Chadian mandible dubbed Abel which is estimated to be between 3 and 3.5 million years old. Abel was also recovered from the desert surface in 1995 and not buried which casts a shadow across all dating.

Despite the enthusiasm of Brunet and others to claim Toumai as an early transitional hominid, it was probably the "vulgar chimp" described by its critics.

Daily Times, September 08, 2008

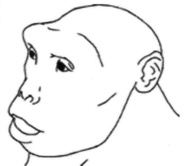

Toumai Man drawing.

http://www.dailytimes.com.pk/default.asp?page=2008%5C09%5C0
8%5Cstory_8-9-2008_pg6_4

ORRORIN TUGENSIS Belongs to the tribe of Homininae which includes humans, chimpanzees and their extinct ancestors. These fossils were found in the Tugen Hills of Kenya by a team led by Brigitte Senut and Martin Pickford from the Muséum national d'histoire naturelle in 2000 and are dated between 6.1 and 5.8 million years old.

The fossils come from five individuals including a femur which suggests that Orrorin walked upright, a humerus shaft, suggestive of tree climbing skills but not brachiation, and teeth suggesting it ate mostly fruit, vegetables and some meat. Senut, one of its discoverers wrote, "The femora indicate that the Lukeino hominid was a biped when on the ground, while its humerus and manual phalanx show that it possessed some arboreal adaptations. The upper central incisor is large and robust, the upper canine is large of a hominid and retains a narrow and shallow anterior groove, the lower fourth premolar is ape-like, with offset roots and oblique crown, and the molars are relatively small, with thick enamel."

http://cogweb.ucla.edu/ep/Orrorin.html

Orrorin's femur is surprisingly modern and morphologically closer to *Homo sapiens* than that of Lucy which came 3 million years later. This of course leads to speculation that Orrorin is not an ancestor of Lucy but a direct human ancestor and that Lucy was merely a cousin. However, there is some debate over the morphology of the femur and its implications.

The team of archaeologists concluded that Orrorin belongs to the hominid lineage which was already present 6 million years ago. In this model, they believe the divergence between apes and humans

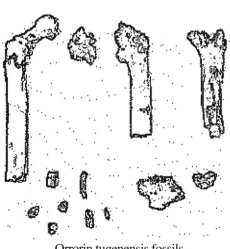

Orrorin tugenensis fossils.

took place prior to 6 million years ago, probably between 9 and 7 million years ago. Australopithecus is an extinct line while *Ardipithecus ramidus* is the ancestor of Pan and the apes.

ARDIPITHECUS RAMIDUS was first unearthed in the remote Afar desert of Ethiopia in 1992 by Gen Suwa from the University of Tokyo. Within two years more fossils had been discovered and were studied in intense secrecy by the Middle Awash Project at the Rift Valley Research Service in Abbis Ababa. Years of fieldwork uncovered the hominin's skull, teeth, arms, hands legs, pelvis and feet, all which had to be painstakingly reconstructed. The skull itself had been crushed into over 60 pieces which were scattered about and has been extensively studied by an international team lead by anthropologist Tim White. This hominin lived 4.4 million years ago and shares human and apelike characteristics.

Since 1992 bones from about 36 members of this species have been studied. 'Ardi' was a 110 pound adult female less than four feet tall who walked upright and was flatfooted unlike knuckle walking apes. Her feet had large opposable toes which allowed her to climb and swing through the trees with ease.

The significance of Ardi is huge for evolutionary scientists. In October 2009 it was announced that humans did not evolve from knuckle walking chimpanzees as had been believed. On the contrary, Ardi shows that humans evolved along a separate lineage from the last common ancestor shared by early hominids and extinct apes. This pushes back the search for the last common ancestor to about 6 million years ago.

Ardi had the head of an ape but could walk upright and didn't use her arms for walking. She retains a primitive big toe that could grasp branches.

Not surprisingly, criticisms soon emerged not only of the purported habitat of Ardi, but whether she was indeed a hominid. Esteban E. Sarmiento of the Human Evolution

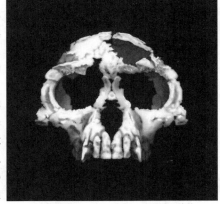

Ardipithicus ramidus, or "Ardi."

Foundation in East Brunswick, N.J challenged the identification of Ardi as a hominid. In a one page article in 'Science' he cited the skeleton's primitive aspects which "predate the human/African ape divergence." Richard Klein, a Stanford University anthropologist said in an interview, "I frankly don't think Ardi was a homind, or bipedal."

In his rebuttal, Dr.White's team responded that Dr.Sarimento was basing his argument on the hominid-ape divergence at 3 to 5 million years ago, whereas recent discoveries have pushed the divergence to 6 million years ago. http://www.nytimes.com/2010/05/28/science/28fossil.html?ref=science

Reconstruction of "Ardi."

Other scientists don't believe that Ardi lived in open savannah grassland but rather a closed canopy woodland. Grassy plains have long been considered the favored habitat of early prehumans and perhaps accounted for their transition to bipedalism.

KENYANTHROPUS PLATYOPS was discovered in Lake Turkana, Kenya in 1999 by Justus Erus, who was part of Meave Leakey's team. Dated from 3.5 to 3.2 million years old, this hominin has a broad flat face with teeth showing both human and ape features. Nicknamed 'Flat face' it is the only fossil found (KNM-WT-40000) and may in fact be a distorted *Australopithecus afarensis*. Other researchers speculate its flatter face is similar to KNM ER 1470 *'Homo rudolfensis'* and it may be a direct ancestor. Flat face has a small earhole and small brain like chimpanzees, but also shares hominin features such as high cheek bones and a flat plane beneath its nose. More than 30 skull and tooth

Kenyanthropus platyops.

fragments were discovered at the site.

Dr. Leakey believes that this specimen belongs to a new genus of ancestors and is the oldest "reasonably complete" cranium found to date. It was a contemporary of *Australopithecus afarensis.*

HOMO GAUTENGENSIS

A new human ancestor, *Homo gautengensis,* which emerged over 2 million years ago and died out approximately 600,000 years ago, was described in a paper accepted for publication in 'HOMO-Journal of Comparative Human Biology' in May 2010.

Darren Curnoe, an anthropologist from the University of New South Wales School of Biological, Earth and Environmental Sciences, led the project which studied partial skulls, jaws, teeth and other bones found at various times at South Africa's Sterkfontein Caves. With colleague Phillip Tobias, a South African palaeoanthropologist, Curnoe believes the new species stood just over 3 feet tall (1 meter) and weighed about 110 pounds (50kg). Although bipedal, it "probably spent considerable time in trees, perhaps feeding, sleeping and escaping predators," he said.

The classification of H*omo gautengensis* upsets the latest theory, proposed by De Ruiter of Texas A & M University that *Africanus sediba* was the 'transitional' species between *Australopithecus africanus* and *Homo erectus* because A. sediba was "Much more primitive than *H. gautengensis,* and lived at the same time and in the same place," according to Curnoe. This makes *"Australopithecus sediba* look even less likely to be the ancestor of humans." Curnoe proposes that *Australopithecus garhi* which lived about 2.5 million years ago in Ethiopia is our earliest non-Homo direct ancestor.

http://www.msnbc.msn.com/id/36177857/ns/technology_and_science-science/

REFERENCES:

Wikipedia references and images on Kenyathropus

'Monkey or man? Toumai, hailed as our oldest ancestor, is stirring ancient scientific rivalries,' James Meek, 'The Guardian,' Oct 10, 2002.

http://www.whyevolution.com/toumai.html

Daily Times, September 8, 2008

http://www.dailytimes.com.pk/default.asp?page=2008%5C09%5C0

8%5Cstory_8-9-2008_pg6_4

'Orrorin Tugensensis: Pushing back the hominin line,'
http://cogweb.ucla.edu/ep/Orrorin.html

'The Oldest known prehuman revealed,' David Perlman, San Francisco Chronicle, October 1, 2009.
http://www.sfgate.com/cgi-bin/article.cgi?f=/c/a/2009/10/01/ BA4K19U6IR.DTL

'Tree-swinger may be earliest human,' Jennifer Viegas, 'Discovery News,' May 21, 2010.
http://www.msnbc.msn.com/id/36177857/ns/technology_and_ science-science/

'Scientists Challenge "Breakthrough" on Fossil Skeleton' New York Times, John Noble Wilford, May 27, 2010.
http://www.nytimes.com/2010/05/28/science/28fossil. html?ref=science

PART 3
OUT OF PLACE and TIME
SKELETAL REMAINS

**ANATOMICALLY MODERN HUMANS IN
ANCIENT STRATA?
EXTREMELY ANOMALOUS CLAIMS —
PRE TERTIARY AND TERTIARY FOSSILS?
TWO MILLION YEAR OLD HOMININS
IN CHINA?
ANCIENT RACES OF THE AMERICAS
AUSTRALIAN PREHISTORIC RACES
PRE MAORI RACES OF NEW ZEALAND
THE TAKLAMAKAN MUMMIES—
CAUCASIANS IN PREHISTORIC CHINA**

ANATOMICALLY MODERN HUMANS IN ANCIENT STRATA?

The following section examines claims of anatomically modern remains found in very ancient strata which have been either ignored or rejected by traditional palaeontologists. The reason is simple- such skeletons should not be found in very early strata because according to Darwinian evolution, they are not supposed to exist. Some, indeed have been proven to be intrusive burials. On the other hand, some are true marvels which cannot be explained away. Unfortunately most fossils in this category are redated or ignored. Orthodox Darwinism must be preserved at all costs. The majority of these examples come from Cremo's 'The Hidden History of the Human Race.'

AFRICA

RECK'S SKELETON

While the oldest hominid skeletons come from Africa, there is also evidence that modern humans could have been living there over one million years ago. Professor Hans Reck of Berlin University was the first scientist to conduct investigations at Olduvai Gorge in Tanzania, then German East Africa.

In 1913 one of Reck's African collectors found parts of a complete modern skeleton embedded in solid rock. Reck had the skeleton removed by chipping it from the rock with hammers and chisels.

Reck identified five stratigraphic beds at Olduvai Gorge and the skeleton was from the upper part of Bed 11, which is 1.15 million years old. Aware of the implications of this ancient find, Reck considered and dismissed the possibility that the skeleton had arrived in Bed 11 through burial.

Eventually Reck convinced Louis Leakey that the skeleton had been found in Bed 11, but in 1932 zoologists C. Foster of Cambridge and D. Watson of the University of London claimed that the completeness of the skeleton clearly indicated it was a recent burial. Furthermore, they believed that red and yellow sediments which formed the bottom of upper Bed 111 had been found in the original

106

matrix, indicating the site was contaminated by a later burial.

Professor Boswell, a geologist from the Imperial College in England claimed that he had been sent samples from the original matrix containing the colored pebbles from Bed 111. This was puzzling because neither Reck nor Leakey had observed such pebbles during their examinations. Reck and Leakey eventually were forced to retract their initial observations concluding that "it seems highly probable that the skeleton was intrusive into Bed 11 and that the date of the intrusion is not earlier than the great unconformity which separates Bed V from the lower series." (Cremo p 236)

Even though Reck's skeleton was redated to Bed V, it is still controversial as the base of this bed is about 400,000 years old, much older than anatomically modern humans are supposed to be. In his 1935 book 'The Stone Age Races of Kenya' Leakey again revised the age of Reck's skeleton to about 10,000 years old, the same age as those found at Gamble's Cave.

Although the skeleton disappeared during the war its skull and bone fragments remained which were supposedly radiocarbon dated in 1974 to 16,920 years. There are some problems with this. For instance the bone fragments were much smaller than usual to obtain carbon and the remains could easily have been contaminated in the 60 years since its discovery. Cremo believes such elements conspired to deliver a falsely young age for the Reck skeleton. He wrote, "Concerning the radiocarbon tests on Reck's skeleton reported by Protsch, the laboratories that performed them could not have dated each amino acid separately. This requires a dating technique (accelerator mass spectrometry) that was not in use in the early 1970s…We can only conclude that the radiocarbon date Protsch gave for Reck's skeleton is unreliable. In particular, the date could very well be falsely young." (ibid p238)

The completeness of the skeleton and its contracted position indicate Oldoway man is a more recent intrusive burial rather than a natural fossil, a position eventually taken by Reck himself.

JAWS, SKULLS, FEMORA AND HUMERI

Various skeletal remains have been discovered where the dating has been controversial. In 1932 Louis Leakey discovered a jaw at Kanam, Kenya and skulls from nearby Kanjera which showed modern features. The fragments of five skulls, designated Kanjera 1-5 were found in deposits from 400,000 to 700,000 years old.

The Kanam jawbone with two premolars was recovered from a solid block of travertine dating from at least 2 million years ago. At the time Leakey believed the fossils showed that a hominid resembling modern humans had existed at the same time as Java and Peking man, so that *Homo erectus* could not be a direct human ancestor.

Initially Leakey's conclusions that the remains had been discovered from undisturbed sediment were verified by the Royal Anthropological Institute, but geologist Percy Boswell began to question the age of the Kanam and Kanjera fossils. Although Leakey accompanied Boswell to Africa to resolve his doubts, Boswell submitted a negative report on Kanam and Kanjera to 'Nature' magazine. Boswell claimed that he had not seen the exact site and had found the geological conditions at the sites confused, charges Leakey denied. Boswell's criticisms of the bones prevailed until 1968 when Philip Tobias reopened the question of Kanjera. In a UNESCO conference of September 1969 the Kanjera skulls were unanimously accepted as Middle Pleistocene.

The Kanam jaw has been described as everything from australopithecine (Keith) to *Zinjanthropus* (Leakey) and *Homo sapiens* by Arthur Keith. Cremo wrote, "That over the years scientists have attributed the Kanam jaw to almost every known hominid (*Australopithecus, Australopithecus boisei, Homo habilis,* Neanderthal man, early *Homo sapiens,* and anatomically modern *Homo sapiens*) shows the difficulties involved in properly classifying hominid fossil remains. (p241)

Fluoride, nitrogen and uranium content tests were performed on the Kanam jaw and Kanjera skulls by K. P. Oakley of the British Museum. The results indicated that the bones were "considerably younger" than the other fossils which surrounded them. However, Cremo has noted that when Oakley first published a paper in 1958 discussing the uranium content of the jaw he wrote: "Applied to the Kanjera bones our tests did not show any discrepancy between the human skulls and associated fauna." Cremo believed that Oakley was dissatisfied with these results and later performed additional tests which obtained the results he desired.

Cremo concluded, "All in all, the results of chemical and radiometric tests do not eliminate the possibility that the Kanam and Kanjera human fossils are contemporary with their accompanying faunas. The Kanjera skulls, said to be anatomically modern, would thus be equivalent in age to Olduvai Bed IV, which is 400,000 to 700,000 years old. The taxonomic status of the Kanam Jaw is uncertain...(but) if it is as old as the Kanam fauna, which is older than Olduvai Gorge Bed 1, then the Kanam mandible would be over 1.9 million years old." (Cremo p 244 ibid)

In 1965 Bryan Patterson and W. Howells found a modern hominid humerus at Kanapoi, Kenya (KP 271) which came from a deposit about 4.5 million years old. According to Patterson and Howells the humerus was different from that of gorillas, chimps and australopithecines but similar to modern humans. Henry McHenry and Robert Corruccini of the University of California concluded, "The Kanapoi humerus is barely distinguishable from modern Homo" and "shows the early emergence of a Homo like elbow in every subtle detail." (Cremo ibid p 252)

Creationist Lubenow also supports the designation that KP 271 is human, and claims that the lower humerus is "relatively easy to discriminate between humans and other primates." This creationist claim is one of their strongest arguments because it has not been disproven.

However, because of its date, KP 271 is assigned to the australopithecines, particularly *A. anamensis*. Lague and Jungers who studied the lower humeri of apes, humans and hominoid fossils using multivariate analysis, claim that KP 271 lies well outside the range of human specimens. They conclude: "The specimen is therefore reasonably attributable to *A. Anamensis*, although the results of this study indicate that the Kanapoi specimen is not much more 'human-like than any of the other australopithecine fossils, despite prior conclusions to the contrary."

http://www.talkorigins.org/faqs/homs/a_anomaly. html#kp271

In 1977 a similar humerus was found at Gombore, Ethiopia which has been dated to about 1.5 million years. It was said, by Brigitte Senut, that the Gombore humerus "Cannot be differentiated from a typical modern human."

SOUTH AMERICA

According to conventional thinking humans did not enter South America, via North America and Asia, until about 12,000 years ago, with the possibility of up to 20,000 years ago. However, every now and then anomalous discoveries crop up which challenge such dating. In the 1880s an atlas, the topmost bone of the spinal column, was discovered in an Early Pliocene deposit aged about 3-5 million years old in the Montehermosan formation of Argentina.

Scientists eventually studied the bone and came up with differing conclusions. Palaeontologist Florentino Ameghino accepted it was truly Pliocene and belonged to an apelike human ancestor while Ales Hrdlincka of the Smithsonian Institute demonstrated that the bone was modern and thus an intrusion into a more ancient stratum. Hrdlincka felt the atlas was worthy of being "dropped of necessity into obscurity" and thus any evidence of modern humans living in Argentina over 3 million years ago was actively suppressed.

A human skull was discovered in Buenos Aires by workmen in 1896 eleven meters below the bed of the river Plate. Eventually it came to the attention of Florentino Ameghino who believed it belonged to a Pliocene ancestor of *Homo sapiens*. He called it *Diprothomo platensis*. Ales Hrdlincka initially dismissed it as a modern skull discovered in stratum of 1 to 1.5 million years old, but later thought it was too modern to rule out any great age for it.

In 1921 a human mandible with two molars was discovered in the Late Pliocene Chapadmalalan formation at Miramar, Argentina by Lorenzo Parodi, a museum collector. It was reported to be firmly embedded in the strata which were 2-3 million years old. However, the fact that the molars appeared to be identical to modern human molars ensured that the bones were assigned to a more recent date.

Brazil also has its anomalous skeletal remains. In 1970 a skullcap resembling *Homo erectus* was discovered in a Brazilian museum by Canadian archaeologist Alan Bryan which had been originally recovered from a cave in the Lagoa Santa region.

Critics were quick to point out that the skullcap belonged to an Old World specimen or was a fake. Bryan insisted it was a genuine fossil from Brazil only to have it 'disappear' from the museum soon after.

CASTENEDOLO SKELETONS

In 1860 Professor Guiseppe Ragazzoni, a geologist from the Technical Institute of Brescia, found a human cranium, limbs and thorax at Castenedolo in Pliocene strata. When his colleagues expressed the opinion that they were recent, he threw the bones away. However, he never lost interest and in 1879 other bones were discovered in the same area by Carlo Germani who had purchased the land. "All of them were completely covered with and penetrated by the clay and small fragments of coral and shells, which removed any suspicion that the bones were those of persons buried in graves, and on the contrary confirmed the fact of their transport by the waves of the sea." (Cremo p 136)

A complete female skeleton was found in the middle of a layer of blue clay over 1 meter thick which showed no sign of disturbance. The layer belongs to the Middle Pliocene, about 3-4 million years old.

Professor Sergi, an anatomist from the University of Rome, studied the bones and the site at Castenedolo. He concurred that it was unlikely that the skeleton had been a burial but rather that the bones had come to rest on the shallow sea bottom where they had been scattered by the action of water.

A further skeleton discovered in 1880 was determined by Sergi, Ragazzoni and A. Issel to be a recent intrusion into the Pliocene layers. However, in his own paper, Issel also concluded that all the skeletons had been recent intrusions, a claim Sergi later refuted, but the damage had been done and the Pliocene discoveries were disregarded.

In 1980 K. Cakley conducted chemical and radiometric tests on the Castenedolo bones and declared that they had a nitrogen content similar to that of Late Pleistocene and Holocene Italian sites, making them recent. However, they had an unexpected high concentration of uranium, consistent with great age. To complicate matters further a carbon 14 test yielded an age of 958 for some of the skeletons although they could have easily been contaminated after lying for 90 years in a museum.

Cremo summarized, "The case of Castenedolo demonstrates the shortcomings of the methodology employed by palaeontologists. The initial attribution of a Pliocene age to the discoveries of 1860 and 1880 appears justified. The finds were made by a trained geologist, G. Ragazzoni, who carefully observed the stratigraphy of the site. He especially searched for signs of intrusive burial and observed none. Ragazzoni duly reported his findings to his fellow scientists in scientific journals. But because the remains were modern in morphology they came under intense scrutiny." (p40)

SAVONA SKELETON Another Pliocene skeleton (3-4 my) was discovered in the 1850s at Savona, Italy at the bottom of a 10 feet deep trench. Arthur Issel declared that the Savona human "was contemporary with the strata in which he was found." His contemporaries were more critical and believed it was a recent intrusion. Father Deo Gratias, who had been present at the excavation, gave a report at the International Congress of Prehistoric Anthropology and Archaeology at Bologna in 1871, indicating it was not an intrusive burial. He stated, "Had it been a burial we would expect to find the upper layers mixed with the lower. The upper layers contain white quartzite sands. The result of mixing would have been the definite lightening of a closely circumscribed region of the Pliocene clay sufficient to cause some doubts in the spectators that it was genuinely ancient, as they affirmed. The biggest and smallest cavities of the human bones are filled with compacted Pliocene clay. This could only have happened when the clay was in a muddy consistency, during the Pliocene times." (Cremo, p 141)

FOXHALL JAW was discovered in England in 1855 at a depth of 16 feet in a quarry. The condition of the jaw, filled with iron oxide, was consistent with incorporation in this bed. This level is about 2.5 million years old. English scientists including Charles Lyell and Thomas Huxley were sceptical of this age because the shape of the mandible was not archaic. It was also unfossilized. The Foxhall jaw disappeared in the early 20th century and is not considered ancient by any modern authorities.

IPSWICH SKELETON An anatomically modern skeleton was discovered beneath a layer of glacial boulder clay near Ipswich, England in 1911. The skeleton was found at 4.5 feet in deposits as old as 450,000 years by Reid Moir. Aware that it could have been a recent burial, he carefully verified the unbroken and undisturbed nature of the surrounding strata. However, when stone tools of the Aurignacian era, about 30,000 years ago, where found nearby, he was forced to change his mind.

Cremo and Thompson wrote, "In Moir's statements we find nothing that compels us to accept a recent age of 30,000 years for the skeleton. Sophisticated stone tools, comparable to those of Aurignacian Europe, turn up all over the world in very distant times….Therefore we cannot agree with Moir that the discovery of tools of advanced type at the same level as the Ipswich skeleton was sufficient reason to reinterpret the site stratigraphy to bring the age of the skeleton into harmony with the supposed age of the tools." (p130)

CLICHY SKELETON was discovered 5.25 meters (17.3 feet) beneath the surface of a quarry on the Avenue de Clichy, France in 1868 by Eugene Bertrand. Sir Arthur Keith believed that they were found in a similar layer to the Galley Hill skeleton, making it approximately 330,000 years old.

Claims of fraud were levelled immediately when a workman was said to have stashed the skeleton in the pit, but a number of scientists still believed the find was genuine. Professor Hamy said, "Mr Bertrand's discovery seems to be so much less debatable in that it is not the first of this kind at Avenue de Clichy. Indeed, our esteemed colleague, Mr Reboux, found in that same locality, and almost at the same depth (4.20 meters), human bones that he has given me to study." (Cremo p128) Bertrand also told the Anthropological Society that he had found a human ulna in the same stratum, although it had crumbled to dust.

GALLEY HILL SKELETON In 1868 workmen discovered a skeleton deeply embedded above a thick bed of chalk at Galley Hill, near London. The skeleton appeared to be of a modern human found in deposits dating to 330,000 years ago. Despite stratigraphic evidence, K. Oakley and M. Montagu concluded in 1949 that the skeleton must have been recently buried in the Middle Pleistocene, particularly as its nitrogen content was similar to fairly recent bones. However, as the Galley Hill bones were found in loam, it is possible that this sediment which preserves protein could have been responsible for the high nitrogen content.

The skeleton's fluorine content and carbon 14 confirmed that it was about 3,310 years old although Cremo was quick to point out that these tests were unreliable because of contamination in the laboratory.

MOULIN QUIGNON JAW In 1863 an anatomically modern human jaw was discovered by J. Boucher de Perthes in the Moulin Quignon pit at Abbeville, France. Found with Acheulean tools at 16.5 feet deep, it was dated in deposits to be 330,000 years old. Further excavations in the presence of trained scientific observers yielded more bone fragments and teeth.

Allegations of fraud soon followed, first with the tools and later with the jaw itself. The jaw had a coloring "which was found to be superficial," although British anthropologist Sir Arthur Keith said this feature "Does not invalidate its authenticity." French anthropologists believed in its authenticity until the belief that Neanderthal man represented a Pleistocene phase in the evolution of modern races made such a find impossible to accept.

NORTH AMERICA

Several extremely anomalous human skeletons have allegedly been recovered from North America, particularly the Sierra Nevada Mountains of California.

THE CALAVERAS SKULL In 1866 a skull was removed from a layer of gravel 130 feet below the surface near Angels Creek, Calaveras County. It is very possible that it was found in deposits older than the Pliocene.

The encrusted skull eventually was examined by J.D Whitney, the state geologist who affirmed it was found in Pliocene strata.

114

Soon, however, it was being denounced as a fake by the religious press and Smithsonian Institution. William Holmes of the Smithsonian examined the skull and concluded it "Was never carried and broken in a Tertiary torrent, that it never came from the old gravels in the Mattison mine, and that it doesn't not in any way represent a Tertiary race of man."

The Calaveras skull.

(Cremo, p 144) Other scientists who studied the skull agreed that the attached gravel was not from the mine but from a cave burial.

However, there were dissentions. W. Ayres, writing in the 'American Naturalist' in 1882 stated: "I saw it and examined it carefully at the time when it first reached Professor Whitney's hands. It was not only incrusted with sand and gravel, but its cavities were crowded with the same material; and that material was of a peculiar sort…" Ayres believed it was the gold-bearing gravel found in mines, not a recent cave deposit

In 1928 Sir Arthur Keith said: "The story of the Calaveras skull…cannot be passed over. It is the 'bogey' which haunts the student of early man… taxing the powers of belief of every expert almost to the breaking point." (Cremo ibid p145)

More recent anthropologists have no doubt that the Calaveras skull was a fake. According to data presented by Taylor (1992) and Dexter (1986), a radiocarbon date of 1,000 B.P. has been recovered from bones found with it. Apparently one of the principal participants in the discovery of the skull admitted to J. M Boutwell that it he was involved in the hoax.

Despite charges of being a hoax, this skull does possess heavy brow ridges which are indicative of ancient hominins like *Homo erectus* or Neanderthals although they have, supposedly never been found in the Americas.

CHAPALA BASIN A skull fragment found at the Chapala Basin, Mexico was found amongst the skeletal remains of over 500,000 specimens, including antelope, camel, bear and sloth. All the specimens, including those modified by humans, were fossilized

115

and stained with a black manganese derived patina which, according to anthropologist Frederico Solorzarno, can be dated to between 50,000 and 80,000 years old.

Whilst sifting through these remains, Solorzarno came across a thick curved bone, no more than a few inches in length. Surprisingly, it resembled the supra-orbital ridge of a European *Homo erectus*. Furthermore, one of the mandible pieces does not correspond to any modern hominin and may, or may not belong to the same individual.

Solorzarno is a respected researcher and former university lecturer in Guadalajara. He says the brow bone raises "many questions, one of them being its great and amazing resemblance to primitive hominid forms whose presence in the Americas has not been generally accepted."

Other scientists have made guarded comments although efforts to date the fragments have so far failed due to lack of surviving tissue.

"Most people sort of just shook their heads and have been baffled by it," said Robson Bonnichsen, director of the Center for the Study of the First Americans at Texas A&M University. "That doesn't mean it's not real. It just means there's not any comparative evidence." That primitive brow ridge from Lake Chapala "is in a category by itself," Bonnichsen concluded.

The Chapala fragment raises disturbing questions in archaeology circles. Not only is it far older than other American *Homo sapiens* fossils, but it also resembles *Homo erectus*, a species which has never been uncovered in the Americas. Furthermore, *Homo erectus* died out over 100,000 years ago in other parts of the world,

It remains a fossil of great interest and controversy.

'Mexico discovery fuels debate about man's origins,' John Rice, Deseret News, Oct 2, 2004
http://www.deseretnews.com/article/1,5143,595095698,00.html?pg=3

ANOMALOUS FOOTPRINTS IN MEXICO

While creationist footprints such as those recovered from Paluxy can be discounted by archaeologists, other less famous footprints are highly anomalous. Dr. Silvia Gonzales, Professors David Huddart and Matthew Bennett discovered footprints in an abandoned quarry at Toluquilla, Mexico. In 2005 these researchers claimed the footprints were much earlier than 11,000 years old, the age of the Clovis settlements, and dated the volcanic rock at 40,000 years old and hypothesized that early hunters walked across the freshly deposited ash.

However, Paul Renne, director of the Berkeley Geochronology Center and his colleagues in Mexico and at Texas A & M University, reported in the December 1 issue of 'Nature' that the rocks had been dated to 1.3 million years!

This date was unacceptable, leaving Renne with two possibilities: "One is that they are really old hominids- shockingly old- or they're not footprints."

Tim White, palaeoanthropologist at UC Berkeley, dismissed the "so called footprints" because of their age. "The evidence (the British team) has provided in their arguments that these are footprints is not sufficient to convince me they are footprints. The evidence Paul has produced by dating basically means that this argument is over, unless indisputable footprints can be found sealed in the ash."

http://berkeley.edu/news/media/releases/2005/11/30_fp.shtml

Graduate student Joshua Feinberg studied the rock grains in the volcanic ash and discovered their polarity was opposite to the Earth's polarity today, making them between 1.07 and 1.77 million years old. However, as each individual grain in the rock is magnetized in the same direction, it is unlikely that the original ash had been weathered into sand that early humans walked through before the sand was welded into rock again. He asked, "If they were hot, why would anyone be walking on them?" (ibid)

Renne has changed his mind about the footprints. "They're scattered all over, with no more than two or three in a straight line," and because of their extreme age, long before the evolution of modern humans, "We consider such a possibility to be extremely remote."

The British team, on the other hand, led by Gonzalez, claim to have found 250 footprints and also those from other animals. Team member Matthew Bennett of Bournemouth said, "Accounting for the origin of these footprints would require a complete rethink on the timing, route and origin of the first colonization of the Americas."

EXTREMELY ANOMALOUS CLAIMS—
PRE TERTIARY AND TERTIARY FOSSILS?

The Tertiary era lasted from approximately 65 million to 1.8 million years ago and saw the rise of mammals and hominids like the australopithecines. According to traditional palaeontology, no humans or hominids existed in North America at this time. Cremo and Thompson described a few anomalous but dubious discoveries from this period in their book 'The Hidden History of the Human Race.' In July 1857 a skull fragment was dispatched by its discoverer, Dr. Winslow, to the Museum of Natural History Society of Boston. It had been recovered at a depth of 180 feet from a shaft in Table Mountain in a gold drift. A stone mortar was found in the same mine. The skull fragment was found in deposits from 9 million to 55 million years old.

Another skeleton was discovered by Dr. H. Boyce at Clay Hill in El Dorado County, California in 1853. Although firmly cemented together, they began to crumble upon exposure to air. William Sinclair cast doubts on the discovery by stating that he could not locate the same clay stratum and that it was probable that the skeleton was a recent internment.

Even more ancient skeletal remains have been reported in the 19th century. In December 1862 the 'Geologist' journal reported that the bones of a human were discovered in a coal bed, ninety feet below the surface in Macoupin County, Illinois. "The bones, when found, were covered with a crust or coating of hard glossy matter, as black as coal itself, but when scraped away left the bones white and natural." (Cremo p 150) This coal seam, from the Carboniferous period, was between 286 to 320 million years old.

Ed Conrad is a man on a mission to prove that humans inhabited America during the Carboniferous era. However, as he is not a typical creationist, his discoveries may be worthy of discussion. In 1981 he discovered an anomalous 'fossil' while exploring abandoned anthracite surface mining operations near Shenandoah, Pennsylvania. After sending a photo to the Smithsonian Institution, he was invited to bring his skull like fossil in for examination. However, after briefly examining it, the experts from the Smithsonian such as Rye dismissed it as an anthropoid skull and declared it to be a worthless rock.

Undaunted, Conrad took his 'fossil' home and cleaned a cavity, discovering what he considered to be a dental arch with teeth. A photo was taken and forwarded to Wilton Krogman, author of 'The Human Skeleton in Forensic Medicine,' an expert in human comparative anatomy and former professor of anatomy and physical anthropology. Krogman identified it as a premolar tooth possessing a pair of cusps.

On Krogman's recommendation, an infrared scan was performed at the American Medical Laboratories in Fairfax, Virginia in 1981, but no-one, including Krogman was able to interpret the graphic results. Once again Conrad asked the Smithsonian for assistance and Rye offered to do more tests. In the meantime, Conrad had discovered what appeared to be a large cranium in the same area as the jaw.

Conrad examined some granules from the cranium like object in the rock and discovered Haversian canals which can be only found in bone. The Smithsonian examined his original fossil again but their report failed to mention any cell structure and concluded that the material was composed entirely of quartz.

At this point Conrad began to accuse the Smithsonian of acting without integrity and complained to his local congressman. On his website, edconrad.com, Conrad writes,

"Most importantly, the Smithsonian's experts knew that if a human skull was discovered in Carboniferous strata, it means that man inhabited the earth multi-multi-millions of years before Darwin's evolutionists have put him here.

"They also knew -- in one felt (sic) swoop -- it would decimate the evolutionary theory of man's origin from some lowly animals of 60-65 million years ago, since Ed's discovery means man was around long, long before.

"Since established science has long maintained that coal was formed more than 280 million years ago, the Smithsonian was well aware that if it confirmed Ed's discovery, it would shake the very foundation of its most close-vested theories."

An article by Bill O'Brien says:

"There was a time when Conrad regarded the integrity of the scientific establishment as beyond reproach. But after seven years of dealing with paleontologists and archaeologists, he said he has

119

found them to be a devious and untrustworthy bunch whose actions in relation to him have been downright dishonest and deceitful."

"Conrad believes his discovery has frightened members of the archaeological/ paleontological establishment out of their wits. They dread the truth, he says, because they

Ed Conrad and his anomalous fossil.

know their cozy little clique will be gone with the eons. No longer will they be able to sup at the trough of Darwinism, enjoying soft jobs with huge salaries."

In 1996 geologist Andrew MacRae from the University of Calgary examined some of Conrad's 'fossils' and determined that they were natural rock. Biologist Paul Myers of the Temple University in Pennsylvania offered to review Conrad's samples with his high powered microscopes. On the site 'Pharynguala' Myers described Conrad as "not your typical reactionary christian creationist, but an odd breed spawned by weird television pseudoscience, paranoid conspiracy theories, von Daniken and Velikovsky, and a deep hatred of scientific authorities."

Myers compared Conrad's specimen with that of a dinosaur and human bone and claimed that the Haversian canals were basically an optical illusion. However, he did make this rather enigmatic statement: "Scrapings from human bone or from Ed's specimen looked essentially identical. This is not to imply that Ed's specimen was bone, however; no Haversian structure was visible, and basically what we were viewing was granular, rough-edged chunks of mineral, whether bone or stone was irrelevant."

Myers concluded, "The sections of human and fossilized dinosaur bone looked nearly identical in dimension, although I will concede that the Haversian structures were slightly larger in this sample of dinosaur bone. The differences were minor, however. The structures were clear and unambiguous at all levels of magnification used.

"No comparable image of Ed's specimen could be acquired. The structures were very granular, with an irregular, variable density. There were clear spots scattered throughout the specimen that might have the approximate dimensions of the central canals in bone, but none of the surrounding structure was consistent with bone. There was also no evidence that these spaces had any longitudinal continuity, so I'm inclined to accept MacRae's interpretation that these are simply transparent grains of quartz."

The irrepressible Conrad responded with these points: 1. Bone structure varies with the type of bone and position within the bone. 2. EC96-001 (his specimen) is petrified and therefore most of the structure has been destroyed.

A rather exasperated Myers countered: 1. None of the diagnostic anatomical features of human bone are present in any of the specimens Ed Conrad has shown.

2. None of the histological features of bone are present in any of the specimens that have been examined.

3. It completely and unquestionably failed to meet any reasonable histological standard for bone.

On Pharynguala Myers concluded, "A sad, sad case. Conrad is a harmless kook, but it is unfortunate to see such a vivid example of imperturbable irrationality."

The whole debate was on the website:

http://pharyngula.org/index/weblog/comments/blast_from_the_past_a_visit_from_ed_conrad_circa_1996/

Conrad still believes his specimens are fossils and that there is a giant conspiracy to discredit him, despite the fact that the Smithsonian did conduct examinations and did enter into lengthy correspondence with him. His website, edconrad.com, has a summary of his dealings with the Smithsonian and Wilton Krogman as well as photos of what appear to be fossilized bones. His stubborn claims may be eccentric but it is prejudicial to call him a "kook" because of his these unorthodox theories, especially as he had the support of Krogman.

MOROCCO

Another highly anomalous 'skull' was reportedly found in the desert of Tafilalet, Morocco in a marble quarry in 2005 by amateur palaeontologist Mohammed Zarouit. As it was found at the same

121

level as Devonian fossils it could be up to 360 million years old. Its 6.1 cm high and 3.9 cm broad skull looks like a small version of the genus homo. It was called *Homo alaouite*, after Dr.. Alaoui-Abdeklader, radiologist of Moulay Ali Chrif hospital who declared it to be an "authentic skull and not a fabricated object."

So-called *Homo alaouite*.

Zarouit claims that X-ray images performed by Dr. Alaoui Abdeklader, radiologist and director of the Moulay Ali Chrif hospital, confirmed "it is an authentic skull and not a fabricated object." Its skull is the size of an apple with a circumference of 18.4 cm. Nevertheless, its jaw, high forehead, dental formula and the position of the occipital hole (in the middle) are all characteristics of the *Homo species*.

http://www.cryptozoology.com/forum/topic_view_thread. php?tid=14&pid=376364

Not surprisingly, this extremely old object has not been taken seriously by palaeontologists. Its extreme age, the fact that the jaw is attached to the maxilia and overall tiny size are unlike anything previously uncovered, including the hobbits of Flores.

REFERENCES:

Cremo. M, 'The Hidden History of the Human Race'

Talk Origins website **http://www.talkorigins.org/faqs/homs/a_anomaly. html#kp271**

'Mexico discovery fuels debate about man's origins,' John Rice, Deseret News, Oct 2, 2004

http://www.deseretnews.com/article/1,5143,595095698,00.html?pg=3

Ed Conrad's website. **www.edconrad.com.**

Pharynguala.org website

Cryptozoology.com 'Authenticity of tiny Tafilalet skull confimed,' Susan Searight-Martinet'

http://www.cryptozoology.com/forum/ topic_view_thread.php?tid=14&pid=376364

Site miroir du Center d'Etude de Recherche sr la Bipedia Initiale

http://cerbi.ldi5.com/article.php3?id_

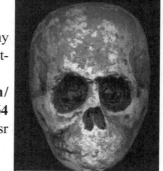

Homo alaouite.

122

article=167
UC Berkeley News, Press release, Robert Sanders, 30 November, 2005
http://berkeley.edu/news/media/releases/2005/11/30_fp.shtml

TWO MILLION YEAR OLD HOMININS
IN CHINA?
OUT OF ASIA HYPOTHESIS REVISITED

China enjoys a rare position in the anthropological world where the emphasis is usually upon African and European hominins. For one thing, it has a long history of palaeontological excavations which extends back to the Peking Man discoveries. Chinese archaeology was also one of the earliest recipients of funding by the Rockefeller Institute. The Chinese are extremely nationalistic and willing to pursue exhaustive research into the evolutionary origins of humanity and the Chinese race.

Various discoveries of *Homo erectus* fossils are fuelling a debate on the origins of our genus *Homo*, while some Chinese scientists openly propose that *H. erectus* evolved separately in China and parallel to that already observed in Africa. Although they don't go as far as claiming that *Homo erectus* and its ancestors originally evolved in China and not Africa as is the current paradigm, they adhere to Wolpoff's 'Multiregional Continuity Theory.'

The site of Renzidong Cave in Anhui Province, eastern China, has yielded animal bones and possible stone tools which indicate that *H. erectus* may have been there 2.25 million years ago. The site is actually a fissure into which many ancient animals fell and were butchered by hominins. No hominin remains were uncovered.

http://www.archaeology.org/0001/newsbriefs/china.html

Discovered in 1984 and excavated from 1985-88, the cave of Longgupo in Sichuan province yielded many fossils, including two hominin teeth, a fragmentary mandible and stone tools. The hominin teeth were originally ascribed to a new species of *Homo erectus, Homo wushanensis*, (Wushan Man) which actually resembled older East African fossils such as *H. ergaster and H. habilis*. Scientists dated the oldest level at 1.96 million years to 1.78 million years.

According to an article by Russell Ciochon in 'Nature' magazine in November 1995, "The Longgupo specimens closely resemble East African fossils representing the earliest species of the genus *Homo*.

They share few characters in common with Asian *Homo erectus*. As their incompleteness precludes designating a new species, we assign the Longgupo hominids to *Homo* species indeterminate, while noting affinities with *H. habilis and H. ergaster*. The stone tools are consistent with this interpretation. Given the early date and primitive morphology for the Longgupo specimens, and the older age estimates for *H. erectus* in Java, we must recognize more than one Plio-Pleistocene hominid species in east Asia. The new evidence suggests that hominids entered Asia before 2 Myr, coincident with the earliest diversification of the genus *Homo* in Africa. Clearly, the first hominid to arrive in Asia was a species other than true H. *erectus*, and one that possessed a stone-based technology. A pre-*erectus* hominid in China as early as 1.9 Myr provides the most likely antecedents for the in situ evolution of *Homo erectus* in Asia."

'Early Homo and associated artefacts from Asia,' Huang Wanpo et.al

http://www.uiowa.edu/~bioanth/nature95.html

Strangely, many anthropologists today are silent on these extremely early dates and the assertion in the article's abstract that "We report here that the hominid dentition and stone tools from Longgupo Cave are comparable in age and morphology with early representatives of the genus *Homo* (H. *habilis* and H. *ergaster*) and the Oldowan technology in East Africa. The Longgupo dentition is demonstrably more primitive than that seen in Asian *Homo erectus*. Longgupo's diverse and well- preserved Plio-Pleistocene fauna of 116 species provides a sensitive contextual base for interpreting the early arrival of the genus *Homo* in Asia." (ibid)

Not only the Out of Africa theory is at stake, but the entire proposition that the earliest hominins evolved only in Africa is also under pressure with these discoveries. Chinese palaeontologists are very cautious and will support the 'Multiregional Continuity Theory' which still alleges an African homeland, rather than assign a totally new area for human evolution. Some Chinese, however believe that humans evolved in China in a direct progression from *Lufengpithecus*, a Miocene ape. Indeed, the Longgupo teeth bear a striking resemblance to this Yunnanese ape and are seen by some as a 'missing link' between hominoids and later Chinese hominins.

There are of course many critics to the theory that Longgupo

124

was a pre erectus hominin. Some believe that the descendants of *Lufengpithecus* were not human ancestors and that Longgupo represents an unknown continuation of the *Lufengpithecus* ape lineage into southern China. Dennis Etler, in his paper 'Implications of New Fossil Material Attributed to Plio-Pleistocene Asian Hominidae' tries to prove that the teeth of Longgupo are "virtually identical" with those of *Lufengpithecus.*

http://www.chineseprehistory.com/art1.htm

However, he does concede the possibility, for the sake of argument, and admits that there could have been three or four distinct hominin lineages in Asia- a gracile Homo in China, *H. modjokertensis* in Java similar to *H. habilis*, larger *H. erectus* in China and Java and a robust australopithecine in China and Java. Or the possibility arises that *H. erectus* actually evolved in Asia from an earlier *Homo* and returned to Africa fully formed. For this theory to be feasible is contingent "on the correct dating, identification and interpretation of the specimens considered…many of the specimens under consideration are very fragmentary and/or very poorly preserved and subject to varying at times contradictory interpretations. The dating of other specimens is still subject to doubt…"

In June 2009 Russell Ciochon published another article in 'Nature': 'The mystery ape of Pleistocene Asia' in which he did a complete turn about regarding the Longgupo fossils. "For many years, I used Longgupo to promote this pre-erectus origin for *H. erectus* finds in Asia. But now, in light of new evidence from across south-east Asia and after a decade of my own field research in Java, I have changed my mind. Not everyone may agree; such classifications are always open to interpretation. But now I am convinced that the Longgupo fossil and others like it do not represent a pre-erectus human, but rather one or more mystery apes indigenous to south-east Asia's Pleistocene primal forest. In contrast*, H. erectus* arrived in Asia about 1.6 million years ago, but steered clear of the forest in pursuit of grassland game. There was no pre-erectus species in southeast Asia after all."

http://www.nature.com/nature/journal/v459/n7249/full/459910a. html

His opinion had changed after he observed 33 primate teeth from Mohui cave in Bubing Basin, south China. Some teeth belonged to Pongo (apes) and others were from *Gigantopithecus*, but 8 were unable to be firmly classified. Ciochon decided that these teeth belonged to a mysterious ape. As for the stone tools found, he dismissed them as recent additions and the Longgupo teeth as an intrusion because they were found on a different level.

It is interesting to study the response of other scientists to Ciochon's defection. Archaeologist Richard Potts does not question Ciochon's reversal, but his reasoning that "through a long history of analyzing fossil sites in Africa where the genus Homo was found, we know that these early humans were living in grassland or savannah-fringe environments. If early humans lived in these more open environments in Africa, why would they inhabit a subtropical forest in Asia?"

Potts cautioned against black-and-white designations like 'forest' or 'grassland' dwellers. "I think," he said, "that we shouldn't let the environment dictate the taxonomy."

http://news.nationalgeographic.com.au/news/2009/06/090617-early-human-ape-mystery_2.html

Considering the fact that early hominins, like current humans, did and do live in rainforest environments, and the fact that tools and modern looking incisor teeth were recovered from the same site, it might be prudent not to dismiss Longgupo man yet.

REFERENCES:

'Early Homo erectus tools in China,' R. Ciochon and Roy Larick, 'Archaeology,' Jan/Feb 2000

http://www.archaeology.org/0001/newsbriefs/china.html

'Early Homo and associated artefacts from Asia,' Huang Wanpo et.al
http://www.uiowa.edu/~bioanth/nature95.html

Dennis Etler, 'Implications of New Fossil Material Attributed to Plio-Pleistocene Asian Hominidae,'
http://www.chineseprehistory.com/art1.htm

http://www.nature.com/nature/journal/v459/n7249/full/459910a.html
'The mystery ape of Pleistocene Asia,' Russell Ciochon, 'Nature,' June 2009

http://news.nationalgeographic.com.au/news/2009/06/090617-early-human-ape-mystery_2.html

ANCIENT RACES OF THE AMERICAS

According to accepted historical thinking, the oldest race to inhabit the Americas was the Clovis people, based upon skeletal remains found in New Mexico. Despite the fact that older remains have been found in Monte Verde, Chile and other sites, the general consensus still remains that the Clovis culture were the original people to cross into the Americas from Siberia via the sunken land bridge of Beringia about 12,000 years ago.

However, the following examples of the earliest skeletons unearthed in the Americas indicate that this area was multiracial. A 2010 study conducted on the earliest skulls dating to 11,000 B.P. reveals a disparity between morphological and genetic studies. Palaeoanthropologists from Brazil, Chile and Germany who studied the skulls came to the conclusion that two separate migrations took place and "These differences are so large that it is highly improbably that the earliest inhabitants of the New World were the direct ancestors of recent Native American populations." However, these scientists were unable to make the paradigm shift away from an Asian origin, despite the fact that these skulls had distinctly non Asian features.

"We conclude that the morphological diversity documented through time in the New World is best accounted for by a model postulating two waves of human expansion into the continent originating in East Asia and entering through Beringia. The disparity between our results and those of most genetic studies points to a large gap in our understanding of the peopling of the New World."
http://www.physorg.com/news195759989.html

A more logical conclusion, perhaps, based upon the morphological measurements of the skeletons, is that various races such as Asian, Polynesian and European were the earliest settlers of the Americas.

PENON WOMAN In 1959 the oldest skeleton ever found on the American continent was unearthed from a prehistoric lake near Mexico City. It consisted of a skull and almost complete skeleton. Scientists, under the leadership of Dr. Silvia Gonzalez from Liverpool's John Moores University and Oxford's Research Laboratory of Archaeology, carbon dated the skull to 13,000 years old in 2002.

127

Steve Connor, Science Editor of 'The Independent' wrote, "However, the most intriguing aspect of the skull is that it is long and narrow and typically Caucasian in appearance, like the heads of white, western Europeans today." (03 December, 2002)

Penon woman's extreme age suggests two scenarios. Either there was a much earlier migration of Caucasian like people with long narrow skulls across the Bering Strait, or more controversially, a group of Pleistocene Europeans crossed the Atlantic at the height of the Ice Age.

Dr. Silvia Gonzalez working at John Moores accepted that her discovery lends weight to the highly contentious idea that the first Americans may have been Europeans. "At the moment it points to that as being likely. They were definitely not Mongoloid in appearance. They were from somewhere else. As to whether they were European, at this point in time we cannot say 'no'."

"My research could have implications for the ancient burial rights of North American Indians because it's quite possible that dolichocephalic man existed in North America well before the native Indians," she said.

Even more controversial is the suggestion that Penon woman could be a descendant of Stone Age Europeans who crossed the Atlantic 15,000 years ago or earlier.

Critics like Professor Chris Stringer at the Natural History Museum in London responded, "Most humans in the world at that time were long headed and it doesn't surprise me that Penon woman at 13,000 years old is also long headed."

Penon woman, like Kennewick and other early American skeletons, certainly brings into question the current belief that Asiatics were the first people to inhabit the Americas.

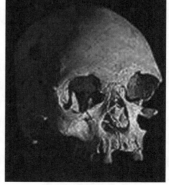

LUCIA According to a BBC News report on August 25, 1999, "The first Americans were descended from Australian aborigines." This astounding claim is based upon a study of the dimensions of ancient skulls found in Lopa Vermelha, Brazil, the second oldest

Skull of "Lucia."

128

in the Americas. Walter Neves, an archaeologist from the University of San Paolo took extensive skull measurements from dozens of skulls, including the oldest, a young woman they named Lucia, who died about 12,000 years ago.

"The measurements show that Lucia was anything but mongoloid," he said. "It has all the features of

Reconstruction of "Lucia."

a negroid face," agreed Richard Neave from the University of Manchester.

The skull dimensions and facial features match closely the native people of Melanesia and Australia but it begs the question how such people could have reached South America about 10,000 years ago. Some archaeologists speculate that an incredible sea voyage, from Australia to Brazil via Polynesia, must have been undertaken by accident.

The skull features changed between 9,000 and 7,000 years ago, from exclusively negroid to exclusively mongoloid. It appears that the mongoloid people from the north invaded and exterminated the original Americans although some of the gene pool may have survived in Terra del Fuego, at the tip of South America.

http://news.bbc.co.uk/2/hi/science/nature/430944.stm

SPIRIT CAVE, NEVADA Discovered in 1940, but only analyzed racially in 1994, these two bodies are the oldest mummies in the Americas. The two bodies had been buried in the cave wrapped in tule matting. Knives, baskets and animal bones were also found, while the second mummy had a complete scalp with a small bit of hair attached. One of the mummies was wearing textiles with a pattern of advanced diamond plaiding.

For four decades the mummy was placed in the Nevada State Museum's storage facility and was assumed to be at most a couple of thousand of years old. However, tests in 1994 revealed it was much older. Using accelerator mass spectrometry, anthropologist E. Erv Taylor submitted 17 samples of bones, hair textiles and wood to reveal that the body was of a 9,400 year old middle aged man of non mongoloid racial origins. Taylor's testing, and that of other

129

A reconstruction of the Spirit Cave, Nevada, skull.

scientists, lead to the undeniable fact that the Spirit Cave Mummy is not an ancestor of any modern Indian tribe. Danise, of the Nevada State Museum, described its "compellingly Caucasoid traits"- a long small face with a large cranium. It resembles either a Japanese Ainu or Norseman from Europe, but not a Mongolian.

However, once these facts were announced the American Indian tribe the Paiutes laid claim to the corpse under the Native American Grave Repatriation Act (NAGPRA) of 1990 which allows for the return and reburial of Native American bodies. They filed a claim of "cultural affiliation" asserting that the Spirit Cave Mummy was a Paiute, despite contrary scientific evidence. Since then they have consistently refused DNA testing of the corpse.

In July 2000 the Bureau of Land Management (BLM) issued a determination that the Spirit Cave Man could not be culturally linked to the claiming Fallon-Paiute Shoshone tribe. The tribe filed a lawsuit asking the Federal Court to review their claim under NAGPRA.

LOVELOCK

Other skeletal remains also indicate that non Asian, possibly European or Pacific Islanders were living in the Americas until about 7000 BCE. A cave near Lovelock, Nevada has yielded several very tall red-haired mummies and up to sixty other skeletons buried under several layers of bat excrement. Discovered in 1911, they

A jaw and skull from the Lovelock Cave compared with a modern human jaw.

purportedly ranged from 6.5 feet to 8 feet tall which is extremely anomalous, although anthropologists claim that the tallest skeleton was only 5ft 11 inches. The local Paiute tribe has a legend of red haired people called the Si-Te-Cah whom their ancestors exterminated. Unfortunately the fate and whereabouts of the red haired mummies is uncertain

OTHER SKELETONS

The Wizard's Beach man, also found in Nevada at Pyramid Lake in 1978, has been dated to 9,225 years old. It has a distinctively long skull shape which is different from the Mongoloid shape of original Amerind inhabitants.

In 2002 it was announced by the Mexico City's Museum of Anthropology that an extremely old skull had been identified as more than 13,000 years old. Called the 'Penon Woman 111', this skeleton has a long face, 'doliocranic' which is not typical of early Amerind skeletons.

Other Caucasoid like skeletons have been repatriated to the Indian tribes and not studied any further. These include:

- Brown's Valley man, Minnesota- 8,900 years old discovered in 1933. His jawbone was as wide as that of *Homo erectus*, and his tools showed a mixture of Yuma and Folsom types.

131

His features resembled an indigenous Greenlander.

- Gordon Creek Woman, Colorado was discovered in 1965 and had facial features more like Europeans or Africans than Asians. Dated to 9,700 years old, no DNA testing has been allowed.
- Pelican Rapids Woman, Minnesota has been dated to 7,800 years. Measurements of 'Minnesota Woman' indicate she is an archaic *Homo sapiens*, being more primitive than the Amerindian or Mongoloid, with a long head. This could also be interpreted to mean that she had Caucasoid features.

http://csasi.org/2000_july_journal/earliest%20americans.htm

KENNEWICK MAN was discovered in July 1996 on the banks of the Columbia River at Kennewick, Washington. The discovery includes a collection of 380 bones and bone fragments, the most complete set of remains ever found in the Northwest.

Kennewick man's long narrow skull and other physical characteristics do not match those of modern Native Americans. Despite its non Indian characteristics, the Colville, Umatilla, Yahama, Nez Perce and Wanapum tribes claimed the remains as those of an ancestor and demanded its reburial. This started a six year legal wrangle between the Army Corps of Engineers, who wanted to turn the bones over to the Indian tribes, and anthropologists, such as Robson Bonnichsen, who wanted to study Kennewick man. In 2000 Interior Secretary Bruce Babbitt ruled that the bones should be handed over to the tribes for reburial, but in 2002 US Magistrate John Jelderks overturned Babbitt and approved research on the bones. Jelderks in his ruling said that the term 'Native American' requires a 'cultural relationship' with a modern tribe to qualify under the grave protection act. He said his review of 22,000 pages of court documents, including scientific reports, produced no evidence to support any cultural link between Kennewick Man and the Northwest tribes.

Bonnichsen said, "That's terrific, I'm delighted. This has been a long process, and we have been convinced from the first that federal law involving these remains was not being followed."

Don Sampson, director of the Columbia Intertribal Fish Commission and former chairman of the Umatilla tribes was bitter. "This country belonged to Native Americans. We have been dispossessed of our own country. Now we are dispossessed of our ancestors. We pursued this because of our religious obligation to provide for our ancestors' remains. We're obligated, religiously and morally, to do that. Only native people are treated this way in America." (Richard Hill, 'The Oregonian' 8-31-2)

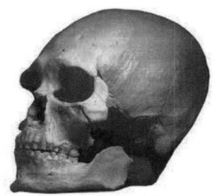

Kennewick Man skull.

Only the Umatillas continued further court proceeding until February 2004 when the United States Court of Appeals for the Ninth Circuit ruled that the cultural link between the tribe and skeletons was not proved, allowing scientific study of the remains to continue. In other words, the tribes failed to prove that Kennewick man was a Native American as defined under NAGPRA.

In July 2005 a team of scientists were given ten days to study the remains, making detailed measurements and determining the cause of death. Initially it was examined by anthropologist James Chatters, who collected 350 pieces of bone, including the skull. A small piece of bone was subjected to radiocarbon dating at the University of California, Riverside and yielded the unexpected date of 9,300 years. After collecting all the pieces, Chatters concluded the skeleton belonged to a Caucasoid male about 68 inches (173 cm) tall.

DNA testing was analyzed but could not be completed because it contradicted Native American values protected under NAGPRA.

Anthropologist Joseph Powell of the University of New Mexico examined the skeleton and came up with a different conclusion. Kennewick man was not Caucasian but resembled southern Asians and the Ainu of Japan.

The remains are now in the Burke Museum at the University of Washington and the controversy continues. Jonathon Marks, Physical Anthropologist, University of California at Berkeley in 1998 summarized the racial controversy and its implications:

133

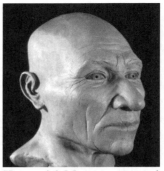

"The 'Caucasoid' Kennewick Man... has been the subject of extensive heated correspondence....To call it 'Caucasoid' is to connote aspects of ancestry, not simply morphology [form and structure]; it directly suggests that America was settled by Europeans and that those now called 'Native Americans' are actually less 'native' than they think. This is a strongly political statement requiring an exceptional level of validation....[The] 'racial' variation in cranial form of prehistoric Native Americans is well attested from earlier studies....Other material similar to Kennewick...also appearing 'Caucasoid' and with a very old date, nevertheless has mitochondrial DNA markers characteristic of American Indians, just as we should expect. So what is the point of racializing these remains? It serves only to clothe 21st-century issues like NAGPRA in the conceptual apparatus and vocabulary of the 19th century."

Kennewick Man reconstructed.

http://www.washington.edu/burkemuseum/kman/idea_of_race.php

REFERENCES:
BBC News report on August 25, 1999, "The first Americans were descended from Australian aborigines." **http://news.bbc.co.uk/2/hi/science/nature/430944.stm**

http://csasi.org/2000_july_journal/earliest%20americans.htm

D. Muska, 'Scalping Science' Nevada Journal, Vol 6, number 2 Feb, 1998

http://nj.npri.org/nj98/02/cover_story.htm

http://www.washington.edu/burkemuseum/kman/idea_of_race.php

'Skulls show New World was settled twice: study' Physorg.com June 14, 2010

http://www.physorg.com/news195759989.html

AUSTRALIAN PREHISTORIC RACES

KOW SWAMP, AUSTRALIA

The Kow Swamp skeletal remains unearthed in Victoria are some of the most unusual *Homo sapiens sapiens* ever discovered. Not only is their age in dispute, but morphologically they resemble earlier hominins which are not officially identified in Australian prehistory.

Archaeological excavations by Alan Thorne at Kow Swamp in the Murray Valley, Victoria, between 1968 and 1972 discovered remains of more than 22 individuals dating from 13,000 B.P. to 6500 B.P. However, optically stimulated luminescence (OSL) tests by Stone and Cupper in 2003 yielded dates from 22,000 to 19,000 years B.P. Unusual features about the Kow Swamp skeletons include 'archaic' characteristics not seen in recent Aboriginal crania but in more ancient *Homo* species such as erectus, or Neanderthals which have never been discovered in Australia.

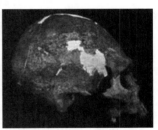

Such features included thicker, more robust skulls, longer with large projecting facial structure. They also had huge cheekbones, big eyebrow ridges, receding foreheads, massive teeth and jaws. Initial descriptions of the skeletons indicated "The survival of *Homo erectus* features in Australia until as recently as 10,000 years ago."

The Kow Swamp skull.

In 1974 Donald Brothwell challenged Thorne's interpretation of the Kow Swamp skeleton by maintaining its archaic shape had been influenced by artificial cranial deformation, particularly in Kow Swamp 5, a claim Thorne has rejected.

In 1977 Thorne and Wilson argued that the Kow Swamp morphological patterns "provide strong evidence that a major morphological changes have occurred in the facial and frontal regions of Aboriginal crania from Northern Victoria over the last 9,000 to 7,000 years."

http://www-personal.une.edu.au/~pbrown3/KowS.html

Another possibility is that the Kow Swamp individuals are representatives of a different race of people, possibly related to those of much earlier Java man. As Mungo Man, the oldest

135

skeleton in Australia (about 35,000 years) is of modern *Homo sapiens* type, this is a distinct possibility. However, this theory is currently out of favor with anthropologists.

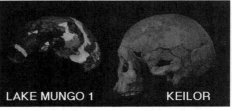

LAKE MUNGO 1   KEILOR

The Kow Swamp skull vs *Homo erectus*.

Similar skeletons were collected near southern NSW at **Coobool Creek** by G.M Black during the 1950s but were not studied until the 1980s. The skeletons were highly mineralized, covered with a thick layer of secondary carbonate and possessing features distinct from recent Aboriginal skeletons. The crania were broken and required prolonged treatment in dilute acetic baths to remove layers of calcium carbonate. Dating was difficult although a pelvic girdle from CC65 obtained a U/Th (uranium/thorium) date of 14,300 (+ or − 1,000 years).

A distinguishing factor of the Coobool Creek and Kow Swamp skeletons is their large crania and marked dolichocephalic nature. The frontal bones are large, with great supraorbital and postorbital breadths. The nasal bones are broad and flattened, indicating a very flat nose. The incisor and canine teeth are large and the skull is thickened.

Other similar robust skeletons have been found in Talgai, Cohuna, Nacurrie and Willandra Lakes, indicating that Kow Swamp and Coobool fossils were not isolated finds. They are referred to as KS (Kow Swamp types) and shared archaic features long after more modern *Homo sapiens* inhabited Australia.

The **Talgai** cranium was discovered in 1884 in the Darling Downs, Queensland, and is heavily mineralized and distorted. Edgeworth David, Professor of Geology at the University of Sydney, who received the skull, persuaded the university to purchase it. S.A Smith wrote the formal description of the skull, claiming it was that of a 14-15 year old male whose cranial vault was "similar in all respects to the cranium of the Australian of today." However, the face and teeth had more archaic characteristics such as large canine teeth. While these features have been refuted by various anthropologists such as Campbell, Burkett, Helman and Macintosh, Macintosh in 1967 emphasized the connection between the Talgai and *Homo erectus,*

136

highlighting its "Prognathic face, its low retreating forehead, and its low vault and huge canine teeth."

Thorne places Talgai, along with Kow Swamp and Cohuna in his robust group of Pleistocene Australians which he distinguished from Lake Mungo types. However, more recent research by Hapgood and Brown found little evidence to support Thorne's two race hypothesis.

Talgai man has not been directly dated but the soil from which it may have originated has been dated to 11,650 years B.P.

http://www-personal.une.edu.au/~pbrown3/Talgai.html

Nancurrie—In 1949 a number of highly mineralized and carbonate encrusted skeletons were excavated by George Murray Black near Swan Hill, Victoria. They were stored at the University of Melbourne until 1971 when Professor Ray, Head of the Anatomy Department, sent them to Professor Macintosh for cleaning, reconstruction and analysis. Eventually one was dated to 11,440 B.P. This Nancurrie fossil possessed a massive mandible and was 180 cm tall.

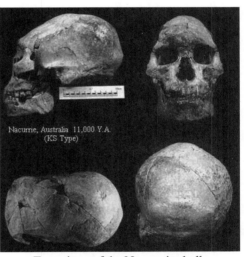

In 1984 a change to the Victorian State Relics Legislation resulted in all of the G. N Black collection, of which Nancurrie was a part, being impounded and reburied.

Four views of the Nancurrie skull.

Willandra Lakes hominid 50 was discovered in the early 80s in south western NSW and the skeleton has not been reliably dated although Caddie in 1987 reported an electron spin resonance date on bone of 29,000 + _ 5000 years).

"The circumstances result in some unease over the extreme claims made about the relevance of WLH 50 to interpretations of the Australasian evolutionary sequence. In particular WLH 50 regularly appears as a corner stone in arguments for evolutionary continuity between the Indonesian and Australian regions published over the last decade," wrote Dr. Peter Brown.

http://www-personal.une.edu.au/~pbrown3/WLH50.html

The skeleton itself has some anomalous features such as a very large cranium and fragments of a very large elbow. The cranial vault is long, broad and high, with a capacity of 1540 ml, as well as some features resembling those of "middle Pleistocene Indonesian hominids" (*H .erectus* such as Java man.)

An interesting comment on the archaic erectus like features of these KS skeletons can be found at **http://www.canovan.com/ HumanOrigin/kow/kowswamp.htm** where Jim Vanhollebeke wrote: "Scientifically, the KS types don't seem to 'fit in' with their primitive features yet recent age. They have remained as odd footnotes in the world of Palaeoanthropology. That their relevance has been ignored is regrettable enough but their rejection as a late chapter to the H. erectus story is unacceptable to this writer. Accepting these fossils for what they are has been a problem for many anthropologists. Part of this problem, possibly, is the fact that the present aboriginal population in this area of the globe, to varying lesser degrees, has been known to exhibit some or all of the traits that make the Kow Swamp type so controversial. This would indicate an obvious line (or lines) of descent...."

In 1997 Jim Foley, creator of the Talk Origins website published his article 'Kow Swamp: Is it *Homo Erectus*?' using the words of Dr. Peter Brown, of the University of New England as his authority. Dr. Brown dissected the 16 stated characteristics of *H. erectus* as listed by creationist author Martin Lubenow and compared the KS fossils to *H. erectus*. Vanhollebeke analyzed his conclusions and his refutations are summarized:

1. *H. erectus skull is low, broad and elongated.* Dr. Brown states the KS crania are not low and ignores the rest of the criteria by adding that these skulls fall within the modern aboriginal range.

2. *H. erectus cranial capacity of 700-1250ml.* Brown only mentions WLH50 which has a brain capacity of 1500 ml and was 'pathological' and Coobool Creek (1404 ml,) ignoring the low scores of Talgai (1300 ml) and Cohuna (1000 ml.)

3. *H. erectus median sagittal ridge* (top of skull middle ridge.) Brown commented that this still occurs in some modern male aborigines and is not diagnostic of *H. erectus.*

4. *H. erectus has suprorbital (brow) ridges.* Dr. Brown claims the Australian samples have only moderate supraorbital ridging and are nothing like *H .erectus.*

5. *H. erectus, Postorbital (behind the eyes) constriction.* Dr. Brown concedes this to KS specimens but qualifies it with the comment that the constriction is not outside the range of recent prehistoric aborigines.

6. *H. erectus, receding frontal contour (sloping forehead).* Dr. Brown pointed out that modern aboriginal crania are receding and the KS or Coobool crania may have been deformed by head binding.

7. *H. erectus occipital bun or torus* (horizontal ridge along back of skull.) Dr. Brown concedes this is common among KS males but "the morphology of the occipital region is nothing like *H. erectus.*"

8. *H. erectus, nuchal area (neck muscle attachment) extended for muscle development.* Dr. Brown admits there is limited development but it is not significant as they all fall within the range of recent prehistoric Aborigines in this respect.

9. *H. erectus, cranial wall unusually thick.* He acknowledges this but dismisses it as being of a different pattern from Pithecanthropus.

10. *H. erectus braincase narrower than the zygomatic arch.* Dr. Brown allows this feature as it is to be expected in "a doliocephalic vault with a well developed masticatory system."

11. *H. erectus, heavy facial architecture.* Dr. Brown wrote "Not what I would describe as heavy."

12. *H. erectus, alveolar (maxilla) prognathism.* Dr. Brown asserted, "big teeth, big palates –prognathic face. The general evolutionary trend has been for a reduction in masticatory system architecture for the last 100,000 years."

13. *H. erectus , large jaw, wide ramus.* Dr. Brown wrote, "large mandible due to large teeth." He also said the wide ramus was lacking in KS.

14. *H. erectus, no chin.* Current aboriginals have a weak mandible. According to Brown, "This is what you would expect with larger teeth and greater alveolar development."

15. *H. erectus, very large teeth.* Dr. Brown concedes this but points out that tooth pattern size is not like *H. erectus.*

16. *H. erectus, post-cranial bones heavy and thick.* Dr. Brown states this is not the case in KS specimens and this is apparently correct.

Vanhollebeke wrote, "The fossil record in Australia shows that there were two basic human populations in late Pleistocene Australia: A. The older (yet more modern and gracile) group represented by excavations such as Lake Mungo. B. The more recent (Yet more primitive and robust) group represented by the KS type examples."

The original arguments can be seen on these websites:

w.talkorigins. http://wworg/faqs/homs/kowswamp.html

http://www.canovan.com/HumanOrigin/kow/kowswamp.htm

Vanhollebeke's photos illustrate the superficial similarity between Homo erectus crania and those of KS variety.

On his **website http://www-personal.une.edu.au/~pbrown3/WLH50.html** Dr. James does invite the reader to compare these photos of Homo erectus and Kow Swamp crania and draw their own conclusions.

Australian naturalist and amateur archaeologist Rex Gilroy believes that *Homo erectus* inhabited the Australian continent, and exists today in isolated pockets as the yowie, a mythical hominoid which wanders the bush.

He believes they entered Australia from Java across an ancient land bridge, although none supposedly ever joined Australia to Indonesia. In his book 'Mysterious Australia' he wrote, "The evidence thus implies that the robust Kow Swamp race are descendants of Java Man *(Homo erectus)* of 500,000 years ago, while the smaller Lake Mungo race entered Australia from China and were probably descendants of the Peking Man (*Homo pekinensis)* and a later Java type, Wadjak Man (*Homo wadjakensis*)."

Gilroy is probably basing this assumption on the incorrect assessment by Creationists that the modern Wadjuk skulls were found in the same layer and place as *Homo erectus* by Dubois. In fact they were found 65 miles away and are of entirely different ages- the modern featured Wadjak man is only about 10,000 years old. Furthermore, Gilroy relies on an artificial land bridge between Indonesia and Australia which has never been established, as well as a fanciful journey from China of Peking men who were, incidentally, also *Homo erectus.*

In conclusion, even though the theory of two Pleistocene races in Australia is now out of favor, possibly for political reasons, the evidence does seem to indicate that the KS robust hominids were markedly different from the older, but more gracile Mungo man.

MUNGO MAN

Three skeletons unearthed in the dried lake bed of Lake Mungo, New South Wales, are the oldest ever found in Australia. However, in contravention of evolutionary theory, they are also much more modern looking than the later Kow Swamp fossils. Lake Mungo 3 (LM3) has been dated between 68,000 and 40,000 years and is one of the world's earliest cremations. LM1, which was poorly preserved, has been carbon dated from 26,000 to 20,000 years old. Its bones were repatriated to the aboriginal tribes in 1992 and reburied.

LM3 is the most interesting of the Mungo burials. His appearance is striking- he was 190 cm or 6 ft 5 in tall and of the gracile type, in contrast with later Kow Swamp people who were more robust. Its age is also disputed and was originally dated between 28,000 and 32,000 years old by palaeoanthropologists from the Australian National University who used stratigraphic comparison with LM1. In 1987 ESR tests established an estimate of his age at 31,000 years, plus or minus 7,000 years. Alan Thorne, using Uranium-thorium dating, electronic spin dating and optically stimulated luminescence dating of the remains and soil came up with an estimate of 62,000 + or − 6,000 years. Since 2003 the general consensus by several Australian palaeoanthropologists is that LM3 is 40,000 years old.

It is the DNA study of LM3 which has ignited the most controversy. In 1995 ANU graduate student Greg Adcock, under the supervision of Thorne and other biologists, began the meticulous process of retrieving DNA from LM3. Not only did he manage to extract mitochondrial DNA from LM3, but was able to compare it with that of nine other ancient Australians who died between 8,000 and 15,000 years ago.

One of three Mungo Men.

ANU evolutionary geneticist Simon Easteal compared the mtDNA with modern aborigines, and thousands of other contemporary groups from around the world, as well as Neanderthals, chimps and bonobos. The results were astounding: LM3, which physically resembled modern humans, had a unique DNA lineage which is now extinct. In other words, LM3's DNA bore no similarity to other ancient Australian skeletons, modern aborigines, Europeans or Neanderthals.

LM3's unique DNA soon became the center of an academic argument between Thorne, an advocate of the 'Multiregional' hypothesis and Colin Groves, an 'Out of Africa' advocate. Mungo man upsets the prevailing Out of Africa theory because it cannot explain how he looked like modern humans but was not related to any humans who had left Africa in the last 200,000 years.

The Multiregional Hypothesis favored by Thorne allows for the possibility that *Homo erectus* could have crossed into Australia from Indonesia anywhere up to 850,000 years ago. Thorne uncovered the Kow Swamp fossils which show some *Homo erectus* features, but lived long after erectus was supposedly extinct. This could suggest that erectus lived in Australia quite recently, or came in a migration wave after *Homo sapiens* and then died out or interbred. However, no self respecting Australian anthropologist will ever admit that *Homo erectus* lived in Australia.

Dr. Groves also had more personal reasons for arguing against the Multiregional Hypothesis, which proposes multiple migrations. He felt it would have been used to reduce pressure on former Prime Minister John Howard to apologize to mixed race aboriginals known as the Stolen Generations who were removed from their communities from 1900 until the 1970s

He wrote:

> But at the same time as one "pure-race" hypothesis was hitting the dust, another was rising. Ancient Australian

skeletons were being discovered in Victoria and southern New South Wales, and they seemed to show great diversity. None of them were Negritos, Murrayians or Carpentarians, but those from Keilor and Lake Mungo were like modern Aboriginal people, whereas some (not all) of those from Kow Swamp had very flat, sloping foreheads, and some people even likened them to so-called "Java Man", Homo erectus, that had preceded modern humans (Homo sapiens) in the region to the Northwest of Australasia at least as late as 300,000 years ago. Unfortunately, although Alan Thorne, the describer of the Kow Swamp skeletons, never actually said that they were Homo erectus, the idea that an extremely primitive people preceded the present Aboriginal people in Australia, and was eliminated by them, seems to have seeped into some folks' consciousness just like the Negritos did. Negritos or Homo erectus - either way, the Aborigines were not the first possessors of Australia so the land doesn't really belong to them and the whites needn't feel too bad about dispossessing them. Really good fodder, this, for the One Nation Party, and the Prime Minister needn't feel he has to say "sorry".

('Australia for the Australians' by Colin Groves:
http://www.australianhumanitiesreview.org/archive/ Issue-June-2002/groves.html)

Groves's speech smacked far more of political opportunism than scientific honesty and highlights some of the glaring problems in Australian prehistory. He admits that ancient Australian fossils show great diversity, from the robust Kow Swamp specimens to the gracile Lake Mungo, but dismisses this diversity because it could be used by such political groups as the One Nation Party to undermine aboriginal rights.

Another theory which is not entertained by any Australian anthropologist is the possibility that the whole theory of Evolution itself may be specious and that humans may have always lived in Australia. Furthermore, there is a strong possibility is that numerous races existed in the island continent which were either wiped out or became extinct.

REFERENCES:
Peter Brown's Australian & Asian Palaeoanthropology
http://www-personal.une.edu.au/~pbrown3/palaeo.html
Rex Gilroy, 'Mysterious Australia'
Talk origins 'Kow Swamp: is it Homo erectus?'
w.talkorigins. http://wworg/faqs/homs/kowswamp.html
J. Vanhollebeke, 'Kow Swamp- is it Homo Erectus? Part 11'
http://www.canovan.com/HumanOrigin/kow/kowswamp.htm
'Mungo Man Turning evolution upside down'
http://www.convictcreations.com/aborigines/prehistory.htm

PRE MAORI RACES OF NEW ZEALAND

The Moriori are the indigenous ethnic group of the Chatham Islands, New Zealand, and many people speculate that they were once prominent on the main islands. They are Polynesian and possibly the descendants of Maoris from the lower South Island who migrated to the Chathams centuries ago.

Unlike the warlike Maori, the Moriori were pacifists and this led to their slaughter by the Maori as late as the 19th century. When about eight hundred Maori armed with guns, clubs and axes chartered the ship 'The Rodney' in 1835, they went on an orgy of killing and cannibalizing the peaceful inhabitants of the Chathams. Those who survived became slaves and the women were forced to have children with their Maori masters. The last full blood Moriori died in 1933.

Current historical thought plays down any evidence of pre-Maori races in New Zealand as Maori land rights are a very sensitive topic. To acknowledge that they were not the original indigenous race of New Zealand but merely another immigrant race would undermine current land claims by the Maori tribes.

And yet according to their own oral traditions, the Maori described numerous other races including those with fair to ruddy skin complexion, blue, green or brown eyes and hair which ranged from fair to black, with red being prominent. Their name for the red headed race was 'uru-kehu'. Other terms were 'Patu-paiarehe,' 'Turehu' and 'Pakepakeha' The wide variety of physical traits indicates that Caucasians, or at least another white race possibly

144

inhabited the islands in prehistory.

Furthermore, according to these legends, there were two at least two distinct races, including pygmies and giants! The pygmies were white skinned, with golden hair and large blue eyes. On the North Island a very tall race of people lived around Port Waikato, the Hokianga Harbour to Mitimiti. They averaged 7 feet tall!

The Maori called these people Pakapakeha, from which the current Maori term Pakeha, used to describe white Europeans, was derived.

Historical references to the smaller races can be found in early journals.

> *'Patu-paiarehe is the name applied by the Maoris to the mysterious forest dwelling race. An atmosphere of mysticism surrounds Maori references to these elusive tribes of the mountains and the bush....The Patu-paiarehe were for the most part of much lighter complexion than the Maoris...their hair was of a dull golden or reddish hue, "uru-kehu", as is sometimes seen amongst the Maoris of today... This class of folk-tales no doubt originated in part in the actual existence of numerous tribes of aborigines. This immeasurably ancient light haired people left a strain of uru-kehu in most ancient tribes'* (see *The Journal of the Polynesian Society*, 1921, volume 30, pp. 96-102, 142-151, article by James Cowan).

The fascinating website, 'Celtic New Zealand,' is run by Martin Doutré who has a keen interest in pre-Maori New Zealand. He has also written a book 'Ancient Celtic New Zealand' which strongly suggests that the pre Maori were of Celtic European extraction. Examples of 'Celtic' influences are stonework and standing stones, art styles and weaving. On this comprehensive site of pre Maori archaeology and mythology, the author writes:

"Sadly, many hundreds of older New Zealand history books, written by on-the-spot witnesses of significant events, or based upon extensive interviews with 19th century & earlier Maori tohungas, etc, are now labeled "Eurocentric-Ethnocentric" and conveniently discounted as unreliable without any rational justification. In the past thirty years or so the dismal works of approved-only, politically-aligned or politically-correct authors have supplanted the comprehensive works of yesteryear. Much subject-matter related to our unravelling New Zealand history or archaeology is now considered unmentionable and not for general public consumption. Scholarship in fields of history, archaeology or physical-anthropology has plummeted to being little more than thinly disguised propaganda. Fortunately, there are still enough old-timers around who have not succumbed to the enforced amnesia programmes that began to rear their ugly heads by the mid-seventies. Many learned Maori elders know from their own oral-traditions that modern selective, sanitised New Zealand history is, nowadays, severely "truth-challenged"... (to use politically-correct terminology)."

http://www.celticnz.co.nz/Bes%20&%20Thor/Bes&Taranis.htm

'Celtic New Zealand' also describes Pre-Maori or non Maori human remains found in New Zealand, particularly in the 19th century. Unfortunately most of these remains no longer exist and are rarely publicized today. In 2004, Member of Parliament, the Honourable Chris Carter, was asked, under an 'official information' request' how many archaeological 'embargoes' were currently in place in New Zealand. In a written response he admitted that there were 105 current embargoes, mostly concerning burial sites, which were classified as 'sensitive records.'

According to 'Celtic New Zealand,' "One of these embargos of recent years included a 75-year suppression of information related to a cache of large stature skeletons at Waikaretu, 12-miles SSE of Port Waikato. The very tall people (measured to be 7-feet or more) were laid out on cut shelves in a cavern, which was exposed during road widening excavations. Anthropologists from Auckland and Waikato Universities were called in and, to the dismay and disgust of the roading contractors, they slapped a moratorium over the find, requiring that it be kept secret from the New Zealand public. Maurice Tyson of Tuakau, a contractor in the area for 50 years, recalls how this upset the men who had discovered the cave. They could not understand why such a valuable, history-changing, archaeological site should be kept secret. In 1988, archaeologist, Michael Taylor slapped an embargo on any release of information concerning the ancient, stacked stone structures in the Waipoua Forest, but that embargo was partially broken by a private citizen's legal challenge after 8-years.

"What this means is that the "powers that be" assume the authority to veto any mention or release of information they consider not suitable for the public. The reality is, however, that all skeletal remains of the pre-Maori people, when located in caves, rock shelters, sand dunes, etc., by hunters or others and reported to the authorities, are inevitably buried, removed or destroyed by concealment teams associated with the local iwi or Department of Conservation. Since the beginning of New Zealand's colonial era, innumerable anomalous skeletons have been seen in dry burial caves and some of these were in coffins or more-often laid out on stone shelves, etc. On rare occasions, some bodies have been seen to be encased within solidified tree gum. Many skeletons have been observed to have the blond, red or brown hair hues, typical of Europeans, and are often accompanied by carved greenstone or other kinds of funerary objects. In coastal sand dunes, as elsewhere, the skeletons are mostly found to be buried in a foetal or sitting position, with the knees drawn up to the chest and trussed (tied). This is similar to Beaker-People burials of ancient Britain or the innumerable mummy-bag burials of Peru." (ibid)

The Maoris paid great respect to the bones of their dead, and yet many remains were found uncovered and untended across the country. During the 1860s to 70s Robertson's mill in Onehunga, Auckland, ground up thousands of these skeletons to make fertilizer because the Maori leaders had told Governor Bowen to "Do with them what you wish, for these are not our people." (ibid)

Even to this day, pre Maori skeletons are being whisked away for burial without being studied. In 2005 12 skeletons were removed from Manutahi farm and re-interred at a Maori burial ground. According to non official sources the remains belonged to ancient people who had no affiliation with the local Maori and some had been buried in a swamp. No photography or forensic analysis of the remains was allowed even though they were in a good state of preservation.

Photos of alleged pre-Maoris taken by early anthropologists show traits

Red haired Moari mummy.

which resemble those of Europeans than Polynesians.

This image from Phillip Houghton's book, 'The First New Zealanders,' compares the physiology of Maori-Polynesians with Europeans. Doutré added the lines and numbers, with 1 being a typical European and 2 a typical Maori. Houghton compared the two:

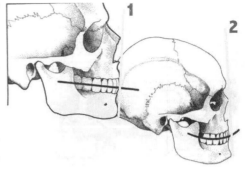

Comparison of Polynesian and European skulls.

- The Polynesian skull is more angular and the European skull more rounded.
- The temples are flat with the Polynesian, with zygomatic arches protruding from the cranial contour, whereas in other races they are usually concealed by the cranial contour.
- Polynesian faces are flat and vertical in profile without any projection of the front teeth.
- The Polynesian cranial vault is high.
- The Polynesian mandible shows a convex curve from front to back which has been named the 'rocker jaw' by anthropologists. (Doutré, ibid)

Many of these so called pre Maoris were buried in tree hewn coffins and placed in high cliff-face tombs. Doutré wrote, 'These skeletons display recognizable European Physiology. They were already very old when found in isolated country, far from the consecrated ground of a churchyard. The deceased people were, undoubtedly, the white Ngati Hotu, known in local Maori and European folklore to have hidden from the cannibals in this inhospitable region." (ibid)

If these pre Maoris existed, and if they were of a non Polynesian race, there is certainly no evidence to link them with any ancient culture in Europe such as the Celts or Vikings. They remain enigmas, as do the Caucasoid skeletons from the Americas.

REFERENCE:

M. Doutré 'Celtic New Zealand,' **http://www.celticnz.co.nz/**

148

THE XINJIANG MUMMIES— CAUCASIANS IN ANCIENT CHINA

Over the past half a century numerous well preserved desert mummies have been found in Xinjiang province, western China. They are known as the Taklamakan Mummies, Tarin Basin mummies, Tocharian Mummies, Urumchi Mummies or Cherchen Mummies depending on where they were found. The basin is extremely arid and full of salt, allowing for almost perfect preservation of an ancient race which bore no resemblance to modern Chinese. Dating from 4,000 years to 2,000 years old, these mummies provide evidence of a hitherto unknown contact between China and Europe/Eurasia during the Bronze Age.

The most famous female mummy has been named 'Beauty of Loulan'. Dated to 2000 B.C., she is the oldest mummy discovered from Qawrighul, near Loulan. A few other mummies were also discovered at Loulan, including an 8 year old child wrapped in a piece of patterned wool cloth closed with bone pegs.

Another blond mummy has been nicknamed 'The Lady of Tarim' and was buried about 3,000 years ago in fine embroidered garments of wool and leather along with jewelry and ornaments of gold, silver, jade and onyx.

Unearthed in 1989, another white female has long blond hair,

The "Beauty of Loulan."

149

partly dismembered limbs and gouged out eyes, suggesting she was a sacrificial victim. She is dressed in woven woolen material which is very similar to Celtic cloth.

Cherchen man, discovered in 1999 by Dolkun Kamberi, died about 3,000 years ago. Aged over 50, he had a two inch beard and brownish hair. An imposing figure when alive, he stood 6 feet 6 inches tall and wore finely woven woolens, deerskin boots and colorful socks. His hands were covered with black tattoos. He was buried with two women and a baby as well as ten hats.

The Cherchen lady was also extremely tall at 1.96 meters. Wearing tall boots and woolen garments, her dress was of a brilliant red hue.

Best preserved of the corpses is Yingpan man, a 2,000 year old Caucasian mummy discovered in 1995. A gold foil Greek inspired death mask covered his blond, bearded face while he wore gold embroidered red and maroon wool garments. Aged about 30 when he died, the 'Handsome Man' was two meters tall, (6 ft 6 inches).

Dolkun Kamberi, a Uighur archaeologist who grew up in Xinjiang, excavated many of the famous mummies. From childhood he had been fascinated by tales of unrecorded mummy discoveries from the early 20th century and legends of an ancient fair headed race. His most unforgettable discovery was in 1985 at Cherchen where he was able to excavate five of the several hundred tombs. In one tomb he found the mummified corpse of an infant wrapped in brown wool with her eyes covered with small flat stones. He also discovered Cherchen man and the Cherchen lady.

Dr. Han Kangxin, a physical anthropologist at the Institute of Archaeology in Beijing, examined the Loulan mummies and determined the skulls were definitely of a European type, some with Nordic features.

Cherchen man, along with dozens of other perfectly preserved mummies, lay in an Urumchi museum until Victor Mair, a professor of Chinese studies at the University of Pennsylvania, came across them. Fascinated by their Caucasian features, he wanted to assembly a team of experts to learn about the mummies. But many scholars, both Chinese and Western, were less than enthusiastic about the discovery.

Some westerners were initially reluctant to invoke the

concept of cultural diffusion in relation to the Tarim Basim as the paradigm of independent invention had been adopted by orthodox archaeology. "Archaeologists were often rebuked if they strayed from this new orthodoxy, which arose in part as a reaction to the political imperialism that often ignored or belittled the histories and accomplishments of subject lands. The Chinese, moreover, had long discouraged research on outside cultural influences, believing that the origins of their civilization had been entirely internal and independent."

(New York Times, 'Mummies, Textiles offer Evidence of Europeans in Far East, May 7, 1996.')

The Chinese were at best unenthusiastic and at worst obstructive. They also had no desire to acknowledge a challenge to their claim of historical sovereignty in the oil rich province of Xinjiang. China supports its claim to the restless province with the myth that it has always been part of China, even though the name Xinjiang means 'New territory.' Separatist Uighurs have seized upon the mummies as proof that their homeland is historically distinct from China by adopting the Beauty of Loulan as a nationalist symbol. They trace their descent to Turkic tribes who settled the area in about 800 A.D. although some Uighurs possess racial features, such as fair hair, of the earlier mummy people.

Chinese historian Ji Xianlin, writing a preface to 'Ancient Corpses of Xinjiang' by Wang Binghua and translated by Professor Mair said, "Within China a small group of ethnic separatists have taken advantage to stir up trouble and are acting like buffoons. Some of them even styled themselves the descendants of these ancient 'white people' with the aim of dividing the motherland. But these perverse acts will not succeed."

Furthermore, the Chinese firmly believed that their culture had developed independently with no European influences until the Roman Empire. According to government

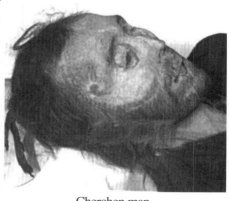

Cherchen man.

151

approved history, China's first contact with westerners which officially dates back to 200 B.C when the emperor Wu Di wanted to establish an alliance with the west against the marauding Huns who were based in Mongolia. The discovery of Caucasian mummies, older than any of Asian appearance, buried with wool, bronze and wagon fragments, all of which were unknown in contemporary China, was highly disconcerting to the Chinese.

Mair had always believed in Trans European contact extending to the Bronze Age and the mummies seemed to prove his thesis. He contacted archaeologist Wang Binghua who had discovered the first mummies in 1978 and was invited to visit a site in 1993. Accompanying them was geneticist Paolo Francalacci from the University of Sassari, Italy. They collected samples from eleven mummies which were still in their graves but the samples were confiscated by the Chinese authorities. However, just before leaving China, a Chinese colleague slipped Mair five of the confiscated samples from two mummies.

Francalacci worked on the samples trying to extract enough DNA for sequencing from the degraded tissues. In 1995 he announced to Mair that preliminary results revealed that the two Xinjiang mummies belonged to the same genetic lineage as most Swedes, Finns, Tuscans, Corsicans and Sardinians.

Mair returned to China where a colleague slipped him another precious gift; a swatch of blue, brown and white cloth taken from a 1200 B.C. mummy. Resembling Celtic plaid, the fabric was shown to be of the exact weave of 13th B.C. salt miners in Austria by Irene Good, textile expert of the University of Pennsylvania Museum.

Mair, Good and fellow textile expert Elizabeth Barber returned to Xinjiang to examine the mummies' clothing. They studied the process of mummification and found it had been totally natural, due to the arid climate and soils.

Barber wrote, "The stacks of clothing buried with the mummies were unlike anything seen before. It just blew me away," she said. "It was like handling 19th century fabric. The first thing that struck me was that it was all sheep's wool, and that really surprised me. I had expected most of it to be plant

Red haired Tarim Basin mummy.

fibre." Sheep are not indigenous to China so these people must have brought them from the west, as they did wheat. Furthermore, the fabric patterns must have been woven on looms similar to those of Europe at that time.

http://www.crystalinks.com/china2.html

After studying linguistic family patterns, Mair attempted to recreate a map of the origins of the Xinjiang mummy race. Their territory stretched from Denmark to the Black Sea, with its heart in Austria-Germany-Czech Republic. After fierce criticism from colleagues, including Barber, he redrew the map, placing their homeland in a broad arc from Ukraine and southern Russia to western Kazakhstan. He published a paper and confessed, "I decided I wouldn't go against the flow that much, because that is a big flow with some really smart people. But in my own integrity and honesty I wanted to put it in here. (Austria)"

Desperately wanting to prove his thesis, Mair returned to Shanghai in 1999 to obtain more samples for a more accurate type of testing. The Chinese were even more difficult and demanded $100,000 for the samples so Mair left empty handed.

'Secrets of the Red-Headed Mummies,' Heather Pringle

http://www.janraisig.com/RedHeadedMummies.pdf

In 2004 a study carried out by Jilin University in China found that the mummies' DNA had a mitochondrial DNA haplotype characteristic of Western Eurasian populations with Eupoid genes. Mair has commented: "My research has shown that in the second millennium B.C., the oldest mummies, like the Loulan Beauty, were the earliest settlers in the Tarim Basin. From the evidence available, we have found that during the first 1,000 years after the Loulan Beauty, the only settlers in the Tarim Basin were Caucasoid."

East Asian people began settling the Tarim Basin about 3,000 years ago. "Modern DNA and ancient DNA show that Uighurs, Kazaks, Krygyzs, the peoples of Central Asia are mixed Caucasian and East Asian. The modern and ancient DNA tell the same story," he added.

The Chinese government finally allowed the project of 'Nova' and England's channel 4 to film a documentary led by archaeologist Jeannine Davis-Kimball.

"They were Caucasoid. This is a no-no for Beijing," she reported of the visit. Her team was closely monitored at all times and even subjected to a hoax planned to mislead them. The team was led to an obviously disturbed tomb which contained a decapitated mummy. She reported, "They had taken the head off so that we would not photograph the Indo-European head."

The Nova program speculates that the mummy people originated in Eastern Europe, near the Black Sea, based upon some nearby petroglyphs which resemble those in Moldova.

'The East Bay Monthly,' VOL XX1X, NO.3 December 1998

The current, politically correct version is that the mummies were of West EuroAsian descent. Mair has commented that the studies were "extremely important because they link up eastern and western Eurasia at a formative stage of civilization (Bronze and early Iron Age) in a much closer way than has ever been done before."

He suggests that these peoples may have arrived in the region by way of the Pamir Mountains about 5,000 years ago and disappeared into the local population by about 800 A.D. when the Uighurs began to dominate. They are also associated with the Tocharian and Iranian branches of the Indo European language family.

Other authorities see the mummies as representatives of Central Asian populations. Han Kangxin examined the skulls of 302 mummies and found the closest relatives to be the Afansevo culture and the Andronovo culture of Kazakhstan. Han in 1998 asserted that the occupants of Alwighul and Kroran were not derived from proto-Europeans but share closest affinities with Eastern Mediterranean populations.

In March 2010 it was announced that the oldest mummies from Small River Cemetery No. 5 had been studied after their excavation in 2003-5. Almost 4,000 years old, they have European features with brown hair and long noses. The cemetery contains a large forest of colored poles or phallic symbols. Beneath each pole were boats laid upside down and covered with cowhide. The bodies inside the boats were wearing large woolen capes with tassels and leather boots and felt caps with feathers tucked in the brim. The women wore skirts made of string strands. Grave goods included grass baskets, masks and bundles of the herb ephedra as well as more phalluses in the female coffins, indicating an extreme interest in fertility and procreation. "The whole cemetery was blanketed with blatant sexual symbolism," wrote Dr. Mair who studied and translated the reports.

http://www.nytimes.com/2010/03/16/science/16archeo.html?ref=scien ce&pagewanted=all

The Chinese geneticists, led by Hui Zhou of Jilin University in Changchun, conducted DNA testing on the mummies, and claimed that the people were of 'mixed' ancestry, having both European and Siberian genetic markers. Co-author, Dr. Jin, had claimed that some of the other mummies, including the Beauty of Loulan, had DNA containing markers indicating an East or South East Asian origin, despite their Caucasian appearance.

The males all had a Y chromosome which is now mostly found in Eastern Europe, Central Asia and Siberia, but rarely in China. The mitochondrial DNA consisted of a lineage from Siberia and two from Europe. Dr. Zhou concluded that the European and Siberian populations probably intermarried before entering the Tarim Basin around 4,000 years ago.

The controversy continues with most authorities downplaying the obvious racial and cultural affiliations between the mummies and Bronze Age Europeans who were proficient at weaving wool, horse riding and building wagons. The Chinese, especially, find this whole topic politically sensitive with the current unrest in Xinjiang.

REFERENCES:
New York Times, 'Mummies, Textiles offer Evidence of Europeans in Far East," May 7, 1996
Nicholas Wade, 'A Host of Mummies, a Forest of Secrets,' New York Times March 15, 2010
http://www.nytimes.com/2010/03/16/science/16archeo.html?ref=scien ce&pagewanted=all
Ancient China, part 2   http://www.crystalinks.com/china2.html
Images-http://www.meshrep.com/PicOfDay/mummies/ mummies.htm

BLOND SALT MUMMIES OF IRAN

Six mummies of varying ages and in various states of decomposition have been recovered from the Chehrabad Salt Mine in the Hamzehlu region of Iran over the past twelve years.

The First Salt Man was accidently discovered in either 1993 or 1994 by miners. According to the 'Tehran Times,' the male was about 35 years old and lived over 1,700 years ago. "He has long white hair and a beard and was discovered wearing leather boots with some tools and a walnut in his possession." However, it is obvious from the photo that the mummy had blond and not white hair which is a very unusual characteristic for both current and ancient Persians.

The Second Salt man, nicknamed Twin Salt Man was found about 50 meters away from the first mummy in 2004. The remains include parts of the skull, jaw, arms, legs as well as hair and nails. The color of its hair was not noted but it was discovered with pieces of wool cloth.

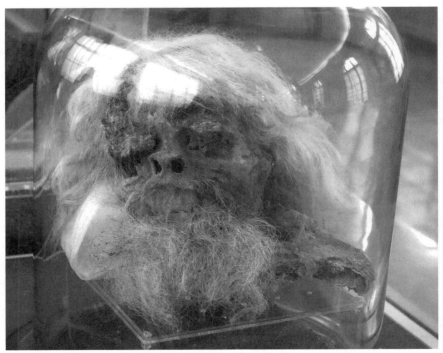

The First Salt Man found in Iran.

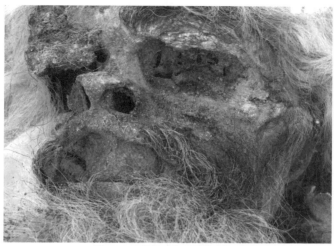
Closeup of the First Salt Man.

The Third Salt Mummy was discovered in 2005 under a large rock that had caused considerable damage to the body. The body was accompanied by items such as a leather sack full of salt, a clay tallow burner, two pairs of leather shoes and two cow horns. Apparently the man had been killed by a rockfall when leaving the mine with a bag of salt.

The Fourth Salt Man was also discovered in 2005 and was the most preserved body to date. X-ray and CT scans indicated it belonged to a 15 or 16 year old male who lived about 2,000 years ago. The boy was wearing two earrings and an iron dagger in a scabbard as well as a knee-length quilted garment and thigh-high leggings.

The Fifth Salt Man was discovered in December 2005 but few details were released.

The Sixth Salt Man was discovered in June 2007 and dates from the Roman era. Apparently it was buried under rocks during an earthquake. This remains in situ.

ANCIENT REDHEADS/BLONDS

Only about 1-2% of the current world's population are redheads, and the largest present day concentration of redheads in the world is in the UK, where up to 40% carry this recessive gene. Redheads can also be found in Northern Europe, Russia, the Basques and their

157

descendants in the New World.

It is surprising to find that red headed mummies can be found all over the world, from Peru to Egypt and New Zealand!

Quite a few Egyptian mummies which have been uncovered have blond or red hair, including the earliest. 'Ginger', who died about 5,400 years ago, was naturally preserved and had golden hair with curly locks. New Kingdom's Prime Minister Yuya's bright

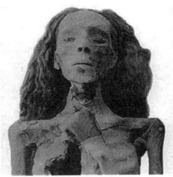

The red-haired Queen Tiy.

blond hair was probably caused by henna dye. The pharaoh Rameses II died at 87 and his mummy had white hair. When the body was sent to Paris to be studied in 1975, Professor P. Ceccaldi studied the hair samples and concluded his hair had been dyed with henna, although the hair roots contained natural red pigments. It is likely that Ramses had red or auburn colored wavy hair and white skin.

Other Egyptian mummies with blond hair belong to Queen Hetop-Heres of the Fourth Dynasty (reddish blond), Queen Hatshepsut, Queen Tiy, daughter of Yuya and mother of Akhenaton.

Pictures of these can be found **at http://guywhite.wordpress. com/2009/01/08/egyptian-mummies-with-blond-brown-and-red-hair/**

Some experts try to explain the hair colors by claiming that the eumelanin in black hair often has broken down in mummies and turns into red pheomalanin when oxidized. However, Joann Fletcher, a leading expert on Egyptian mummy remains, has extensively studied their hair and disagrees. In a paper 'Secrets of the Locks Revealed' Vol. 10, 1998 she wrote, "Hair color is a fascinating study in itself, and the wide range of shades portrayed in Egyptian art does, to a large extent, reflect the diverse range found in reality. The most common hair color then, as now, was a very dark brown, almost black color, although natural auburn and even (rather surprisingly) blond hair are also to be found. With their great fondness for elaboration, the Egyptians' skillful use of dyes has produced yet further shades for us to study, analysis showing many to be various forms of henna, which even an aged Rameses II had used regularly to rejuvenate his white hair."

The red-haired mummy found at Machu Picchu, Peru.

Fletcher also revealed the race of the mummies discovered from excavations at Hierakonpolis. "The vast majority of hair samples discovered at the site were cynotrichous (Caucasian) in type as opposed to heliotrichous (Negroid), a feature which is standard through dynastic times . . ."

http://www.egyptorigins.org/ginger.htm

The pre Inca mummies of Paracas in Peru were considerably taller than other Indian races and had narrow facial features with auburn colored hair. Many were found buried in a trussed, sitting position with their personal possessions, like the earliest New Zealand mummies. This leads to the possibility that some of the pre-Maori population of New Zealand could have migrated by boat from South America.

According to Heyderhal, the chestnut colored hair was, "silky and wavy, as found amongst Europeans, they have long skulls and remarkably tall bodies. Hair experts have shown by microscopic analysis, that the red hair has all the characteristics that ordinarily distinguish a Nordic hair type from that of Mongols or American Indians."

(Heyerdahl "Aku-Aku: The Secret of Easter Island" (George Allen & Unwin, London 1988) pages 351, 352.

159

AMERINDIANS IN EUROPE

In 2007 Norwegian archaeologists reported that a 1,000 year old skeleton at St Nicolas Church is Sarsborg appeared to be Incan. They also discovered the remains of two older men and a baby. One of the skulls had characteristics indicating he was an Inca.

"There is a bone in the neck that hasn't grown and this is an inherited characteristic only found among Inca Indians in Peru. This is sensational," declared Mona Buckholm, archaeologist at the Borgaryssel Museum.

http://www.aftenposten.no/english/local/article1856505.ece

Iceland also has genetic links with ancient America. On November 16, 2010 Spanish scientist Carles Lalueza-Fox of Pompeu Fabra University in Spain, announced that a genetic study of about 80 Icelanders revealed they possessed genes corresponding to an Amerindian who was probably taken from America by Vikings about a thousand years ago. The genetic research undertaken by Spain's Center for Scientific Research will be published in the American Journal of Physical Anthropology.

http://www.telegraph.co.uk/science/science-news/8138884/First-Americans-reached-Europe-five-centuries-before-Columbus-voyages.html

PART 4
STRANGE SKULLS

DOLICHOCEPHALOIDS (CONEHEADS)
PUMPKIN HEAD, M HEAD
HORNED SKULLS
ADENA SKULL
BOSKOP SKULLS
'STARCHILD'

Some of these weird skulls found from various parts of the world, frankly defy explanation.

DOLICHOCEPHALOIDS

A dolichocephaloid skull is by definition long and narrow as opposed to a brachycephaloid skull which is broader and rounder. Historically, dolichocephalic skulls generally belong to Caucasian skeletons and brachiocephalic skulls belong to Asiatic skeletons although there are variations. In recent years extremely dolichocephalic, conical skulls have been discovered in South American museums which defy explanation. It is well known that many ancient cultures practiced head binding which distorted skull shapes but extreme dolichocephalism has rarely been studied by anthropologists.

Cone headed skulls were first mentioned in 1851 in the book 'Peruvian Antiquities' by Mariano Rivero and John James von Tschudi. Dr. Tschudi, with credentials in philosophy, medicine and surgery, described dolichocephalism in two distinct Peruvian races which existed before the Incas, the Aymares and Huancas. The Huancas had the most pronounced dolichocephalic traits although Tschudi had little historical data on them. The Aymaraes had intermediate dolichocephaly.

Even at that time scientists were proclaiming that the skulls had been artificially elongated by head binding, a claim currently in favor for the Australian Kow Swamp skeletons. However Dr. Tschudi commented, "Two crania (both of children scarcely a years old), had in all respects, the same form as those of adults. We ourselves have observed the same fact in many mummies of children of tender age…The same formation of the head presents itself in children yet unborn, and of this truth we have had convincing proof in sight of a foetus enclosed in the womb of a mummy of a pregnant woman… aged 7 months!"

Researcher Lumir Janku has studied many of the anomalous skulls from the Paracas region and divided them into 3 types; premodern, Conehead 1, 2 and 3. The 'premodern' skull has some pre Neanderthal features such as pronounced brow ridges, robust lower jaw and occipital ridge on the bottom back of the skull. Its massive cranial

FRONT AND SIDE VIEW.

Peruvian child mummy.

Two coneheads from the Anthropology Museum in Ica, Peru.

arch, to Janku, suggests "that the skull belongs to a representative of an unknown premodern or humanoid type." His illustration indicates that the brain of the 'premodern' was of a similar capacity to modern humans, despite its extreme elongation.

The Cone heads, as evidenced from three different specimens, C1, 2 and 3 may have developed from the 'premodern' type but reveal a much larger brain.

Dolichocephalic skulls have also been discovered in some of the earliest Old World cultures of Malta and Iraq. Janku wrote, "The enormity of the cranial vault is obvious from all three pictures. By interpolation, we can estimate the minimum cranial capacity at 2200 ccm, but the value can be as high as 2500 ccm." This can be compared to modern humans whose brain capacity is, on average about 1450 ccm.

He believes it is a distinct branch of the genus Homo, if not an entirely different species.

http://www.world-mysteries.com/sar_6.htm#Deformed%20Anci ent%20Skulls

Discoveries in the Maltese temple tombs at Tarxien, Ggantja and Hal Saflieni have yielded three classes of skull anomalies. Much of the research on Maltese skulls comes from the Italian writer Adriano Forgione who classified three types:

163

1. Highly dolichocephalic, "above all, strange lengthened skull, bigger and more peculiar than the others, lacking of the median knitting or suture, linking bones in the roof of the skull."

2. Skulls which were more 'natural' in appearance yet "still presented pronounced, natural dolichocephalous' shapes 'distinctive of an actual race.'

3. 3. Many of the 7,000 skeletons dug out of the Hypogeum exhibit "artificially performed deformities."

Maltese archaeologists Dr. Mark Mifsud and Anthony Buonanno told Forgione, "They are another race although C14 or DNA exams haven't yet been performed."

Forgione complained that the skulls had been removed from display at the Maltese Archaeological Museum of Valletta in 1985. About 15 years later he was given permission when Michael Refalo, the Minister for Tourism, accompanied him to the museum and obtained the director's cooperation. A few days later the Museum's archaeologist Mark Anthony Mifsud placed the skulls before

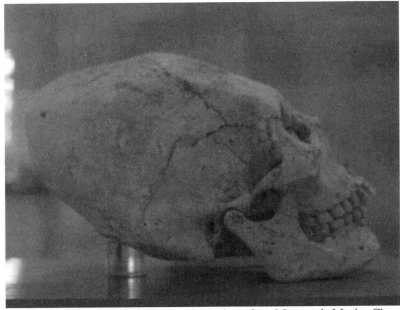

Dolichocephalic skull of an Olmec at the Anthropology Museum in Mexico City.

Forgione, some with pronounced doliochocephalism. They also lacked a median sagitta which is highly anomalous.

In an article for Atlantis Rising Magazine called 'Mystery of Malta's Long-Headed Skulls,' Forgione postulated, "The long head and drawn features must have given an almost serpent-like appearance, stretching the eyes and skin. Lacking the lower part we can only speculate, but the hypothesis seems plausible. Such deformities would certainly have created walking problems, forcing him almost to slither! The lack of the cranium's medium knitting and therefore, the impossibility of the brain's consistent, radial expansion in the skullcap must have caused terrible agony from infancy."

http://www.bibliotecapleyades.net/arqueologia/esp_malta01.htm

The other skulls also showed anomalies, including mild dolichocephalism, indicating that a separate race, distinct from the native populations of Malta and Gozo, existed about 5,000 years ago. They also showed signs of trepanning and artificially performed deformations, such as cranial bondage.

In 1933 Max Mallowan excavated Neolithic graves at Tell

Bust of the Atonist daughter of Akenaton, Meritaton.

165

Chinese drawing of a Taoist conehead.

Arpachiyah in Iraq dated from 4600 B.C. to 4300 B.C. (Halaf and Ubaid periods). He reported the discovery of skulls having a "marked degree of deliberate, artificial deformation", leading to an "elongated skull."

A 1996 monograph on Mallowan's discoveries by Stuart Campbell remarked, "Skull deformation at Arpachiyah appears, on current (1995) knowledge, striking…Skull deformation seems to occur with regularity at other sites of this general period over a

166

very wide area…Jericho, Chalcolithic Byblos, Ganj Dareh, and Ali Kesh."

'The Dolichocephaloids' Randy Koppang **http://www.biped. info/articles/missingrace.html**

Dolichocephaloids also appeared in predynastic Egypt and in the art of the New Kingdom Armana period which covered the era of King Akhenaton. Professor Walter Emery excavated Saqquara in the 1930s and discovered amongst other skeletons a dolichocephalic skull larger than the others. He postulated it was from a different ethnic group which wasn't indigenous to Egypt but had performed priestly and government roles. He associated them with Shemsu Hor, 'the disciples of Horus'.

Forgione believes that the skulls originated with a race who settled in Mesopotamia about 7,000 years ago. In urbanized centers such as Jarmo, the mother goddesses were represented as divinities with the faces of vipers and lengthened heads. They are also associated with the 'Nephilim' of Genesis and eventually settled Egypt in predynastic times. They reached Malta in about 2500 B.C. and survived a millennium later in the mysterious pharaoh of Akhenaton who was always depicted with an extremely elongated skull.

RUSSIA

Pravda News reported on October 6, 2005 that extremely dolichocephalic heads had been discovered in the Caucasus. The Pyatigorsk skull was found near Kislovodsk and dates from between the 3rd and 5th century A.D. "Researchers have repeatedly proved that the skulls had been deformed on purpose," said Dr. Kuznestov. "Ropes or special blocks were tied tightly round the heads of infants, over the temples. The custom went out of fashion by 17th century. The reason behind the deformation phenomenon is still unknown. It is hard to say whether the methods worked effectively or not since nobody ever conducted scientific experiments regarding the binding of the infants' heads."

Whether this skull was deformed by head binding, it bore a striking resemblance to the

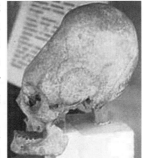

Russian conehead.

skulls of Peru.

In January 2009 it was reported on digitaljournal. com that extreme dolichocephalic skulls had been dug up in Omsk, Siberia by Russian archaeologists. A one minute video purportedly shows skulls dating from the 4th century A.D. which were being studied by scholars at the Omsk Museum of History and Culture.

Tuchersfeld, Germany, conehead.

Igor Skandakov, director of the museum said that the skulls have marks which could be evidence of artificial deformation of a normal skull. He claimed that the skull was kept away from the public because of its unusual shape which shocked and frightened people.

Archaeologist Alexi Matveyev felt that the deformation was carried out as a status symbol of belonging to the elite of society, or as a way to enhance brain function. "It's unlikely that the ancients knew much about neuro-surgery. But it's possible that somehow they were able to develop exceptional brain capabilities."
Digital Journal Feb 28, 2009 -**http://www.digitaljournal.com/ article/268227**

GERMANY

A coneheaded skull is on display in the museum of Tuchersfeld, Germany. The inscription states that about 20 similar male and female skulls were found in the Franconia-Suisse area of Bavaria.
REFERENCES:
http://www.bibliotecapleyades.net/arqueologia/esp_malta01.htm
'The Dolichocephaloids,' Randy Koppang
http://www.biped.info/articles/missingrace.html
Digital Journal, Feb. 28, 2009.
http://www.digitaljournal.com/article/268227

PUMPKIN OR 'J' SKULL

This skull, also from Peru, shows an enormous cranial vault with a brain size of between 2600 ccm and 3200 ccm. Its eye sockets are also about 15% larger than those of modern humans but in most other respects, this skull is of modern size. The age and date of this skull are unknown but it was photographed in a Peruvian museum by Robert Connolly who created a CD of anomalous skulls in 1995. However, author David Childress says that this particular skull is

The Pumpkin skull at the museum in Merida, Mexico.

in a museum in Merida, Mexico. This is probably the real location of this skull.

'M' SKULL

Also part of Connolly's Peruvian collection, this skull has the largest cranial capacity of all the skulls-

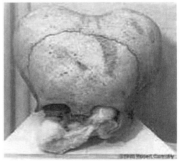

The "M" skull from Peru.

169

Chinese drawing of Taoist gods with "M" skulls.

exceeding 3000 ccm. Unfortunately its mandible is absent so it is difficult to reproduce its features but it is highly anomalous.

Other "M" shaped skulls have been shown in Chinese and Olmec art and sculpture.

HORNED SKULLS

Many gods of the ancient world were depicted with horns as were kings. Horns were a symbol of power and authority. Amazingly, horned skulls have been discovered which are highly anomalous even though sporadic reports of horns growing out of skulls is reported in popular media.

A horned skull was apparently in an exhibit in the Surnateum, Museum of Supernatural History in France. Discovered in the early 20th but now missing, the museum allegedly analyzed the skull and

French horned skull.

170

demonstrated that the horns were part of the skull.

The analysis concluded: 'An in-depth examination and X-rays leave no room for doubt: the skull is not a forgery.'

A website 'Was there a Race of People with Horns?' lists some of the horned skulls which have been discovered. **http://www. bibliotecapleyades.net/vida_alien/alien_watchers04.htm**

These include:

- A burial mound at Sayre, Bradford County, Pennsylvania which was excavated in the 1880s by state historian Dr. G.P Donehoo and professors A. Skinner and W. Morehead. Anatomically normal except for their extreme height, horny projections extended two inches above the eyebrows. The bodies had been buried around 1200 A.D. although they have never been seen again after being sent to the American Investigating Museum in Philadelphia.
- Giant, horned skeletons were unearthed south of Elmira, New York.
- In Texas a male skeleton was unearthed near El Paso with two small horns protruding from the forehead. It was witnessed by a Texas Ranger investigating another murder case.

ADENA SKULLS, OHIO

The Adena were a Midwest tribe which came onto the scene about 3,000 years ago. Their culture extended from Ohio, Indiana, West Virginia, Kentucky to Pennsylvania and New York. The pre-Adena peoples had slender bodies and long heads with narrow skulls. The Adena were far more robust and may have originated in Mexico. They practiced cradleboard head deformation.

The Adena were more short headed, and their cradle boarding supposedly, according to conventional thinking, had the effect of giving the skull a more round dome. In fact Adena skulls have the highest cranial vault found anywhere in the world. They also had prominent foreheads,

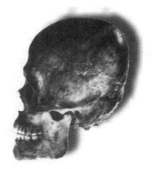

Adena skull.

171

heavy brow ridges, jutting chins and massive bones. Adena folk were very tall with women often over 6 feet and man approaching 7 feet have been found.

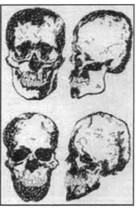

Adena skulls.

So who were these mysterious bulbous headed people who also build mounds for honoring the dead? Their mounds were also used as ceremonial centers and possibly gathering places. DNA testing has found no specific match between the Adena and any existing Native American group. Eventually the Adena people disappeared from the archaeological record and were replaced by the Hopewell culture.

BOSKOP SKULL

The small town of Boskop in South Africa became known to paleontologists in 1913 when two Afrikaaner farmers brought fragments of an ancient skull they had unearthed to the Port Elizabeth Museum. S.H. Haughton, a trained palaeontologist examined the skull and was excited to report to the Royal Society of South Africa that it had a very large cranial capacity of at least 1830 ccm. This is almost 25% larger than the average brain today!

Robert Broom conferred, and reported that the corrected cranial capacity of the Boskop skull was 1,980 ccm, making it larger even than Neanderthal skulls. He named the skull *Homo capensis* and proposed it was related to the European Cro-Magnons. The skull was dated between 30,000 and 10,000 years old.

The facial features of the Boskops were also highly unusual in that they were small and childlike. This retention of juvenile features into adulthood is known as pedomorphic.

However, by the 1950s, the Boskops were demoted from the family tree when paleontologists determined that the larger skulls discovered and named 'Boskop race' were merely variations of southern African tribes.

In 1958 they were resurrected in Loren Eiseley's collection, 'The Immense Journey.' Proposing the idea that the Boskopoids had advanced huge brains and small faces, he suggested that humans,

'Future Man' would eventually evolve into such perfection. This thesis appears to be behind the recent book, 'Big Brain: The Origins and Future of Human Intelligence' by neuroscientists Gary Lynch and Richard Granger.

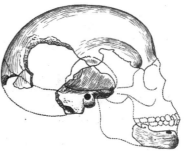

The neuroscientists made some remarkable conclusions about brain size and intelligence according to the

Reconstruction of the Boskop skull.

'Discover' article 'What Happened to the Hominids Who May Have Been Smarter Than Us?' published online December 28, 2009.

They wrote, "Even if brain size accounts for just 10 to 20 percent of an IQ test score, it is possible to conjecture what kind of average scores would be made by a group of people with 30 percent larger brains. We can readily calculate that a population with a mean brain size of 1,750 cc would be expected to have an average IQ of 149.

This is a score that would be labeled at the genius level. And if there was normal variability among Boskops, as among the rest of us, then perhaps 15 to 20 percent of them would be expected to score over 180. In a classroom with 35 big-headed, baby-faced Boskop kids, you would likely encounter five or six with IQ scores at the upper range of what has ever been recorded in human history. The Boskops coexisted with our Homo sapiens forebears. Just as we see the ancient Homo erectus as a savage primitive, Boskop may have viewed us in somewhat the same way."

http://discovermagazine.com/2009/the-brain-2/28-what-happened-to-hominids-who-were-smarter-than-us/article_view?b_start:int=2&-C=

The book has been savaged by palaeontologists such as John Hawks on his blog 'Return of the 'amazing' Boskops. He wrote, 'The portrayal of 'Boskops' in the Discover excerpt is so out of line with anthropology of the last forty years, that I am amazed the magazine printed it. I am unaware of any credible biological anthropologist or archaeologist who would confirm their description of the Boskopoids, except as an obsolete category from the history of anthropology."

Furthermore, he made the valid point that there is no way of

173

estimating the IQ of a fossil skull. To claim that a brain size of 1,750 ccm would correlate with an average IQ of 149, is quite ludicrous and undermines the thesis of the book. But the claim by Granger and Lynch that the large brain also had a prefrontal cortex 53% larger than the average brain today simply cannot be verified.

http://johnhawks.net/weblog/reviews/brain/paleo/return-amazing-boskops-lynch-granger-2009.html

Leaving aside the latest book by Granger and Lynch, it is safe to describe the original Boskop skull as unusually large, but not indicative of any separate race or higher intelligence.

REFERENCES:

'Discover' article 'What Happened to the Hominids Who May Have Been Smarter Than Us?' published online December 28, 2009.

http://discovermagazine.com/2009/the-brain-2/28-what-happened-to-hominids-who-were-smarter-than-us/article_view?b_start:int=2&-C=

.http://johnhawks.net/weblog/reviews/brain/paleo/return-amazing-boskops-lynch-granger-2009.html

'STARCHILD'

Nicknamed 'Starchild' by researcher Lloyd Pye, this skull was reportedly found in a mine tunnel in northern Mexico next to the body of a woman. The shape of the skull is highly anomalous with an enlarged symmetrical cranium and eye sockets close together. Carbon dating shows that the skull is about 900 years old.

The provenance of the skull is questionable although Lloyd Pye received it from a couple in El Paso, Texas in 1999 who claimed it had been recovered from the back of a mine tunnel about 60 to 70 years ago by a teenage Mexican American girl. Pye has a novel theory about human origins and claims that early hominoids such as australopithecines and Neanderthals were not human ancestors. Combining this thesis with the ancient astronaut thesis of Zecharia Sitchin, Pye has constructed an alternative model for the origin of homo sapiens- deliberate interbreeding with extraterrestrials. He believes the skull, which he called 'Starchild', is the product of a human/alien crossbreeding program and has arranged for scientific testing to establish the skull's genetic heritage.

174

Lloyd Pye with his "Starchild" skull.

For months Pye took his skull to various experts, trying to establish if the subject had suffered from a congenital abnormality such as hydrocephalus. In 2000 he handed the skull over to Dr. Ted Robinson, a cranio-facial plastic surgeon from Vancouver, to study. Various studies were performed, such as Carbon 14 and a CAT scan which revealed normal cranial sutures and a complete lack of frontal sinuses. The lower part of the face was missing, but the small chewing muscles indicated that the mandible size would have been greatly reduced from normal. Studies also indicate that the neck muscles were also much weaker, meaning that the neck was 1/3 the size of normal children and centered directly under the mass of the head rather than an inch to the rear.

The 'Starchild's brain was 1600 ccm, much larger than the 1200 ccm for average children. Its eyeball sockets are centered in the middle of the nose, shallow and symmetrical.

175

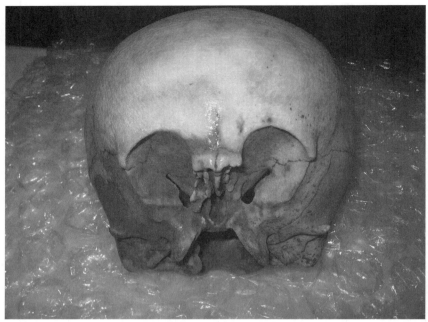

The "Starchild" skull.

Dr. Robinson wrote, "In general, the skull has the basic components of a human skull: ie a frontal bone, two spheniods, two temporal, two parietals and an occipital. However, these bones have been markedly reconfigured from the 'normal' shapes and positions such bones usually have. In addition, the bone itself has been reconstituted to an equally marked degree, being somewhat less than half as thick as a normal human bone, with a corresponding weight of roughly half normal. The reconfigurations and reconstitution are uniform throughout all axes an in all planes of the skull. There is no asymmetrical warping or irregular thinning that is the hallmark of typical human deformity....The morphology of this skull is so highly unusual as to be unique in my forty years of experience as a medical doctor specializing in plastic and reconstructive surgery of the cranium...It seems to be not only unique in my personal experience, but also unique throughout the past history of worldwide study of craniofacial abnormalities."

Regarding the abnormal eyeball sockets he wrote, "If indeed these sockets held eyeballs, those of normal size would have greatly protruded from the face, creating a serious liability of damage during routine activity. Because the eyeballs occupy a position lower in the

176

face than is normal, and they rest in a socket markedly reduced in rectilinear shape and depth, they would have been significantly reduced in size. In either case, however, large eyeballs or small, they would require upper lids three or four times more extensive than normal upper lids to be lubricated in the manner necessary for human eyeballs to function properly."
http://www.starchildproject.com/

The skull's abnormalities can be summarized as thus:
- Skull suturing and baby teeth from the upper palate indicate the child was about 5 years of age.
- The brain volume is about 1600 ccm, in contrast to an average child's volume of about 1200 ccm and an adult's volume of 1400 ccm. Had it lived to adulthood, its brain capacity would have been about 1800 ccm.
- The average human skull weighs about 2.2 pounds. The Starchild skull weighs only 13.5 ounces, although some of it, such as the mandible, is missing. Because it is roughly the same size as a modern adult, the bone must be significantly lighter.
- The skull shows a high degree of symmetry, unlike most cranial deformities.
- A CAT scan has reveal that none of the sutures in the skull have sealed themselves off from further growth making it highly unlikely that the skull is deformed.
- While normal human eye sockets have a recessed conical shape with optic nerves and optic tissue at the rear quadrant, the child's eye sockets have a shallow 3cm scalloped shape with optic nerves and fissures at the inner bottom. The shape and width of the eye orbits are anomalous.
- A recent CAT scan has revealed that the eye canals are larger and have more depth than average human inner ears. We do not know what the external ear looked like.
- The child has small maxillary cheek sinuses but no trace of frontal sinus cavities.
- The Foramen Magnum is positioned to a central point which provides a much better balance between its rear brain, face and forebrain. In average humans the foramen

is positioned slightly rear of the center.

- The child's neck must have been extremely small as determined by a shallow arc extending about 3 cm from the foramen hole, while the inion has become slightly concave. In average humans the neck attachments begin at the inion, the bump in the middle of the occipital bone and sweep into a semicircle that reaches to behind the ears and converges at the foramen magnum. The distance is about 5-6 cms.

- The area available for attaching chewing muscles is reduced as much as the neck muscles. Based upon these muscles, the amount of mandible these muscles could have secured must have been greatly reduced.

- **http://www.bibliotecapleyades.net/esp_starchild_0.htm**

In 1999 initial DNA testing at BOLD Lab, Vancouver, was done on the Starchild and adult female found with it. Although they collected DNA from both skulls, the Starchild proved very difficult to type as the first two attempts produced contaminated samples. On the third try a small sample of 200 pictograms (1000 is normal) proved that it belonged to a male human.

Pye was skeptical that the results could be 100% reliable as the BOLD lab was not designed to recover ancient DNA. He was able to secure funding for a second test at Trace Genetics which were able to process more ancient samples. These tests concluded that both the adult and Starchild had mitochondrial DNA of Native American origin, haplogroups C and A, proving its mother was human.

The Starchildproject discusses recent bone studies which reveal anomalous 'fibers' which have never been observed in any human or other bone. Using a Scanning Electron Microscope (SEM), fibers and a reddish colored 'residue' were found inside the cancellous holes of the Starchild bone. These 'fibers' are so strong that they were not able to be cut by a dremel blade.

In 2003 further DNA analysis was able to recover mitochondrial DNA from the Starchild's Amerindian mother, but could not

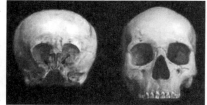

The "Starchild" and a modern skull.

recover any nuclear DNA which comes from both the mother and father. This 'no result' was unable to prove that the father was not human although the suggestion was there.

However, in 2010 there have been many improvements in the recovery process which have yielded amazing results. While some of the nuclear DNA (265 base pairs) matches perfectly with a gene on human chromosome 1,342 base pairs show "no significant similarity" to any genetic material in the NIH database!

Pye wrote in an update, "This past weekend I met with the geneticist working on the Starchild's DNA. He explained how he can now prove the Starchild is not entirely human, which has been our position for years. Now it is no longer a question of 'if,' but of 'when' and 'how' we spread this astounding new reality beyond the mailing list…Please understand this result has now been verified several times and a few more different fragments have been identified that cannot be matched in this database to anything known…I should add that I still can't reveal the name of the geneticist or where he works until we are ready to formally present his results to the world. However, trust me, he is a well-established professional and his facility is large and very credible." **http://exopolitics.blogs.com/exopolitics/2010/03/lloyd-pye-update-we-finally-have-a-recovery-of-nuclear-dna-from-the-starchild-skull-.html**

REFERENCES:
http://www.starchildproject.com/
http://www.bibliotecapleyades.net/esp_starchild_0.htm
http://exopolitics.blogs.com/exopolitics/2010/03/lloyd-pye-update-we-finally-have-a-recovery-of-nuclear-dna-from-the-starchild-skull-.html

STAR-TRIBUNE, Casper, WY - Feb. 1, 2005

Man offers $10,000 for Pedro Mountain Mummy

Hopes to poke holes in theory of evolution with artifact

By BRENDAN BURKE
Star-Tribune staff writer

In the name of poking holes in the theory of evolution, a Syracuse, N.Y., man says he will pay $10,000 for one of the most mysterious artifacts ever dug up in Wyoming — the Pedro Mountain Mummy.

John Adolfi says he wants the Pedro Mountain Mummy, sometimes referred to as Pedro, in order to conduct DNA tests, X-rays, and magnetic resonance imaging on the little fellow.

Conducting such tests, however, is no easy matter, as the mummy vanished in 1950.

Although the mummy has not been seen in public for 55 years,

COURTESY/Casper College Library's David Collection

Bob David holds the Pedro Mountain Mummy in this undated photo. A Syracuse, N.Y., man is offering $10,000 for the mummy that vanished in 1950.

several photos and scriptions of the ar artifact remain.

According to t scriptions and pho the size and propor make the mummy and pique Adolfi's in it.

In the seated p which Pedro is fr stands only 7 inches were to stand up, it mated Pedro would 17 inches. And s standing, he only three-quarters of a

Adding to the Pec tain Mummy enig fact that he is proj much more like an an infant.

According to Ad pothesis, conductin scientific tests on my will reveal that I an adult at the tir death. This would was one of the "littl — a mythical tribe pygmies who Wyoming's mount cording to Arap Shoshone tales.

These little peop postulates, were a s hominid primatos w

PART 5
DWARVES AND LITTLE PEOPLE
OF PREHISTORIC TIMES

PYGMIES OF ANCIENT AMERICA
PEDRO THE MOUNTAIN MUMMY
HOBBITS- HOMO FLORESIENSIS
PALAU PYGMIES

PYGMIES OF ANCIENT AMERICA

The following are reports of pygmy skeletons found in Pre Columbian America. These examples have been provided by William Corliss, a tireless researcher of anomalous science and archaeology.
"Gentlemen's Magazine, 3.8:182, 1837:

"A short distance from Cochocton, Ohio Us, a singular ancient burying ground has lately been discovered. "It is situated," says a writer in Silliman's Journal, "on one of those elevated, gravelly alluvions... From some remains of wood, still apparent in the earth around the bones, the bodies seem all to have been deposited in coffins, and what is still more curious, is the fact that the bodies buried here were generally not more than from three to four and a half feet in length. They are very numerous, and must have been tenants of a considerable city, or their numbers could not have been so great. A large number of graves have been opened, the inmates of which are all of this pygmie race. No metallic or utensils have yet been found to throw light on the period of the nation to which they belonged."

'Anthropological Institute, Journal, 6:100, 1876:

"An ancient graveyard of vast proportions has been found in Coffee County. It is similar to those found in White county and other places in middle Tennessee, but is vastly more extensive, and shows that the race of pygmies who once inhabited this country were very numerous. The writer of the letter says: "Some considerable excitement and curiosity took place a few days since, near Hillsboro, Coffee county, on James Browns farm. A man was ploughing a field which had been cultivated many years, and ploughed up a man's skull and other bones. After making further examination they found that there were about six acres in the graveyard. They were buried in a sitting or standing position. The bones showed that they were a dwarf tribe of people, about three feet high. It is estimated that there were about 75,000 to 100,000 buried there. This shows that this country was inhabited hundreds of years ago."
Unfortunately little has been heard of the pygmy graves

after it was decided that the bones belonged to normal sized children. There have been some dissenters such as V. R Pilapil who hypothesized that the pygmies arrived in America in ancient times from Southeast Asia, probably the Philippines where the diminutive Aetas live. He recalled B Fell's examination of the skeletons, including these facts: 1. The skull capacity was only about 950 ccm, about the volume of a normal sized 7 year old. 2. The teeth showed development and wear of mature individuals.3. The skulls were brachychephalic with projecting jaws, similar to adult Aetas.

Pilapil, Virgilio R.; "Was There a Prehistoric Migration of the Philippine Aetas to America?" <u>Epigraphic Society, Occasional Papers</u>, 20:150, 1991

PEDRO, THE MOUNTAIN MUMMY

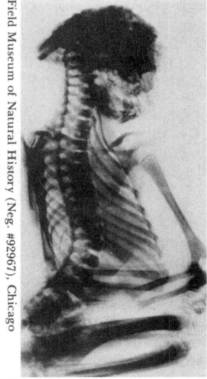

Field Museum of Natural History (Neg. #92967), Chicago

In 1932 a tiny mummy was discovered by gold prospectors blasting the walls of a gulch in the Pedro Mountains, west of Casper, Wyoming. They discovered that the rock face led into a small cavern about 15 feet long and 4 feet high (4.5 m x 1.2) It had been totally sealed from the outside world and inside the cavern was a small ledge upon which the cross-legged creature sat. His hands were folded in his lap, his skin was brown and wrinkled, his nose flat, forehead low, his mouth was broad and thin lipped. Fingernails were preserved and on the top of its head was covered in a dark jelly like substance. The mummy was only 14 inches tall and weighed 12 ounces!

At the time scientists believed

X-Ray of the tiny mummy of "Pedro."

183

"Pedro" the tiny mummy discovered in 1932 in Wyoming, now missing.

Pedro was a mummified pygmy and possibly the progenitor of the American Indian. It was eventually X-rayed by Dr. Harry Shapiro of the American Museum of Natural History and certified as genuine by the Anthropology Department of Harvard University. Pedro had a

184

perfectly formed skeleton with a complete set of ribs and was about 65 years old at the time of death. His spine was damaged, he had a broken collar bone and his skull had been smashed by a heavy blow. The gelatinous substance on his head was exposed brain tissue and congealed blood. Pedro also had the closed fontanels and full set of teeth of an adult, with overly pointed canines.

Unfortunately the mummy disappeared in the 1950s but the X-rays passed on to Dr. George Gill, of the University of Wyoming. The body, he concluded, was that of an infant or fetus suffering from anencephaly which would account for the adult proportions of its skeleton. Dr. Gill hoped to be able to study the original mummy but it still remains missing. It is interesting to note that the Shoshone nation of Wyoming have legends of the Nimerigar, a nasty race of pygmies who kill their infirm with a blow to the head.

HOBBITS - HOMO FLORESIENSIS

In 2003 archaeologists made a remarkable discovery in Liang Bua Cave on the Indonesian island of Flores. An almost complete skeleton named LB1 and complete jawbone of LB2 dated at 18,000 years old were uncovered, including stone tools similar to those of other *Homo sapiens*. The skeleton was of an exceptionally small woman, prompting its discoverers Peter Brown, Michael Morwood and their colleagues to argue that LB1 constituted a new species of hominin, *H. floresiensis*.

The joint Australian-Indonesian team of palaeonanthropologists and archaeologists were digging on Flores for evidence of the original human migration of *Homo sapiens* from Asia to Australia. Subsequent excavations recovered fragments of seven additional diminutive skeletons dating from 38,000 to 13,000 years old. The date is very controversial because prior to the discovery it was assumed that only *Homo sapiens* existed at that time period.

Anatomically *H. floresinsis* is quite distinct from any present or prehistoric hominid, such as the australopithecines.

Homo floresiensis.

185

Not only does it have the body size of a modern three year old, but its cranial capacity was also very small. Less obvious features include the form of the teeth, the absence of a chin and the unusually low twist in the forearm bones.

LB1, nicknamed the Little Lady of Flores stood about 1.06 m (3 ft 6 in) in height, whereas the tibia measurements of another skeleton LB8 suggest it stood about 1.09 meters (3 ft 7 in) which are smaller than the African pygmies or Andamanese Negritos. It is estimated that LB1 weighed about 25 kg (55 lb) making it smaller than not only *Homo erectus* but the African australopithecines.

The skeletons also had a remarkably small brain in the range of about 380 ccm, which is at the lower range of chimpanzees or australopthecines. Furthermore, its brain capacity dwarfs that of its supposed ancestor, *Homo erectus* which creates problems for anthropologists. To try and explain this conundrum, Brown and his colleagues suggested that in the limited food environment on Flores, *H. erectus* underwent strong insular dwarfism. This theory is supported by the discovery of other dwarfed species in Flores such as a dwarf Stegadon, but it has been criticized by Indonesian prehistorians such as Teuku Jacob who argue that LB1 is similar to local Rampasasa island populations.

Professor Jacob, chief paleontologist of the Gadjah Mada University, immediately argued that rather than constituting a new species, LB1 was a microcephalic modern human. "The skull looked to me like a primate's. It was only when I picked it up that I knew it was Homo sapiens. We did the measurements…the shape of the skull from the back is pentagonal. Later I saw the pelvis and the thigh bone. It's just human. It's not erectus….The arm bones, the leg bones…all are small, but that is all. If you analyze the front of the face you might think it is an ape. But look at the whole head and it looks much more human, especially from behind."

In other words, the small skull is from a mentally defective pygmy suffering from the genetic disorder microcephaly which produces a small brain and skull.

Jacob was critical of the Australians' methodology and compared them to "scientific terrorists" who forced ideas on people. He added, "They did their study without comparative material. We are now studying every detail and comparing it with all the other remains from Flores caves and neighboring islands, like the small individuals found in east Java in the 1950s. I have studied the remains from several caves in Flores in the 1960s. There are five similar caves in the area. Catholic priests found some small skeletons in the 1950s. Dutch anthropologists found some in the 1960s."

http://www.guardian.co.uk/science/2005/jan/13/research.science

In December 2004, after the discovery had been published in 'Nature' magazine, Professor Jacob removed most of the remains from Soejono's Institution, Jakarta's National Research Center of Archaeology, apparently without official permission. He eventually returned the remains with portions damaged and two leg bones missing in February 2005 under international pressure. According to reports:

- The chin of the second Hobbit jaw was snapped off and glued back together misaligned at an incorrect angle.

- "The pelvis was smashed, destroying details that reveal body shape, gait and evolutionary history."

- Much of the detail at the base of the skull was removed.

- The left outer eye socket and two teeth were broken off and glued back. Molded rubber was left adhering to some sections.

- Long deep cuts made by a blade used to cut away molded rubber were found on both sides of the jaw.

Professor Morwood was sickened by this wanton act of destruction, but Jacob denied any wrongdoing, claiming that the damages occurred during transport from Yogyakarta to Jakarta.

Fortunately a CT scan had been taken of the skull prior to Jacob's interference and in 2005 a computer-generated model of the skull of *H. floresiensis* and analyzed by a team headed by Dean Falk of Florida State University. The article, published in 'Science' in Feb 2005 claimed that the brain case was not of a pygmy or microcephalic but rather a separate species. However, in October 2005 'Science' published a study headed by Alfred Czarnatzki, Carsten Pusch and Jochen Weber, disproving these findings. Further studies in 'Science'

187

by Martin of the Field Museum in Chicago and the 'Proceedings of the National Academy of Science' also concurred that the skeleton belonged to a microcephalic human.

Morwood and Brown have disputed the 2005 studies, arguing that the skeletons belong to a separate species. Bill Jungers, a morphologist from Stony Brook University examined the skull and found no trace of disease. Debbie Argue of the Australian National University also published a study in the 'Journal of Human Evolution' which concludes that the bones belong to a new species and not a microcephalic human.

In August 2006 Jacob published his findings online in 'Proceedings of the National Academies of Sciences of the United States of America' (PNAS). Not only, he claimed, was the hobbit a microcephalic, but many LB1 traits resemble those of present Austromelanesians, particularly the Rampasasa pygmies of Flores. Hobbits are probably the ancestors of these people and not a separate species, he and his colleagues claimed. Coauthor Etty Indriati also noted that hobbit teeth also share features with Rampasasa pygmy teeth, such as rotation of the premolar.

Ralph Holloway, palaeoanthropologist at Columbia University was dissatisfied with the PNAS paper even though he believed LB1's brain was pathological and assymetrical. He agreed with Brown that the assymetry could result from being under several meters of deposits for thousands as years, particularly as the bones had never fossilized. Holloway, who has also studied an LB1 endocast, believes the brain's size and other features hint at pathology, such as the gyri recti which seems too thin.

In January 2007 Falk published a new study concluding that the bones did not belong to microcephalic humans. She studied 3D computer generated models of 10 normal and 10 microcephalic brains, revealing the floresiensis skulls have a shape more aligned with normal brains but also possessing unique features such as enlarged frontal and temporal lobes and an extended lunate sulcus. "It has an extraordinary morphology unlike anything I've seen in 30 years," Falk told 'New Scientist.' These images revealed a highly developed brain which answered past criticisms that the 'hobbit' brain was too small to be capable of creating the tools which were found nearby. Furthermore, a 2007 study of carpal bones of LB1 found similarities between Australopithecines or chimpanzees. Its shoulders, arms and wrists are close in structure to early hominins or chimpanzees.

Still the separate species critics continue to argue that *H. floresiensis* never existed. In June 2007 Hershkovitz et al. published a paper claiming that the morphological features

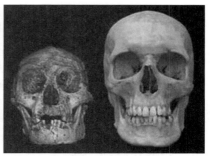

Homo floresiensis and modern man.

of *H. floresiensis* are indistiguishable from those suffering Laron syndrome. Sufferers of this disease exhibit dwarfism, prominent forehead, depressed nasal bridge, under-developed mandible, truncal obesity and a very small penis. However, the vast majority of these reported cases have been in the Mediterranean region, with some in the Bahamas and Turkey, although not Indonesia.

The status of the hobbit today remains questionable, although many people are willing to accept that it belonged to a separate species. An interview with Richard Roberts, an Australian who was one of the leading members of the Flores expedition by the 'National Geographic Channel' responds to some of the criticism.

"A few skeptics (only three to the best of my knowledge) have claimed that the partial skeleton is that of a pathological modern human suffering from an extremely rare medical condition known as microcephaly ("bird-headed dwarfism"). But this is not the case.

"First, we have unearthed the remains of no fewer than seven tiny individuals, so it is stretching credibility to its limits to argue that all of these individuals suffered from microcephaly. Second, the evidence of the two well-preserved mandibles (lower jaws) indicates that the chin and other features are not similar to Australopithecines and Homo erectus, and are not in the Homo sapiens' range of variations.

" In short, the combination of traits in the "Hobbit-like" species' mandible is not present in any Homo sapiens' mandible, pathological or otherwise.

"Third, the proposition becomes even more implausible when the distinctive attributes of the post-cranial skeleton are considered:

189

for example, the pelvis is much wider than in Homo sapiens and the arms are much longer, relative to stature. Neither of these peculiarities is found in modern humans suffering from microcephaly."

The ongoing Hobbit saga raises some controversial issues, even if its authenticity is not established. The role of Dr. Jacob, who virtually hijacked the fossils and allowed them to be damaged, should raise serious questions. He was even accused of washing and dissolving the LB1 skull in acetone to make it impossible to extract any DNA for analysis.

Dr. Jacob, highly regarded as an Indonesian anthropologist, was also known for gathering as many Indonesian fossils and locking them away from further study, unless he approved the research. He was also a supporter of the Multiregionalism evolutionary model which advocates that *Homo erectus* migrated from Africa and spread throughout the rest of the world. His bias was not unnoticed by the Australian team, particularly Roberts who said, "Our team had everyone involved- geomorphologists, geochronologists, archaeologists, paleo-anthropologists...We left no bone unturned." Suggesting that Jacob and his critics had an "intellectual interest" in denying that the skeleton was a new species, he added, "All... are supporters of the multiregionalism evolutionary model...This discovery would destroy their theory. It suits their purposes very nicely (to oppose Homo floresiensis.)"

Before dying in late 2007, Dr. Jacob remained a fierce critic of the *Homo floresiensis* classification and the Australian team who unearthed it. "It is not a new species. It is a sub-species of Homo sapiens classified under the Austrolomelanesid race. If it's not a new species, why should it be given a new name?" Jacob had told a press conference in 2004. "So, if they (the Australian scientists) say the skeleton was the ancestor of the Indonesian people, forget it," he added.

This raises the specter of Nationalism in archaeology which has occurred in other areas of the world. Was Professor Jacob disgruntled that foreign (Australian) scientists had early access to the remains and were thus able to describe them formally in 'Nature' before he was allowed to study them? For several years he blocked further excavations at Liang Bua probably because he did not want to be proven wrong about the hobbit's origins.

190

Jacob's antics also put him at odds with the younger Indonesian paleontologists who are based at the Indonesian Center for Archaeology which is funded by the Australian Research Council.

"We have a big dispute with Professor Jacob," said Tony Djubiantono, chief of the archaeology center and co-leader of the team. "We didn't give him permission to do any of these things."

Independent Hindu Creationist researcher Michael Cremo is also critical that the hobbit constitutes a new species. He wrote in an article, "In *Nature*, the archaeologists proposed that *Homo floresiensis* was the dwarf offspring of *Homo erectus*, who had earlier migrated to Flores Island. But the creature is so different morphologically from *Homo erectus* that its attribution to the genus Homo is to me questionable. For example, pygmy humans have brain sizes somewhat the same as those of normal sized humans, whereas the Flores Island creature has a brain much smaller (in terms of its relation to body size) than *Homo erectus*.

"The discoverers attributed the animal bones and stone tools found along with the bones of *Homo floresiensis* to actions by that creature. But it is possible that the stone tools, which are of an advanced type, and the animal bones, as well as the bones of *Homo floresiensis*, were the work of anatomically modern humans.

He concluded, "So one possible solution to the Flores Island puzzle is that *Homo erectus* apemen from Java crossed the sea and went to Flores Island about 800,000 years ago, and then over hunDr. eds of thousands of years became hobbit like dwarfs. The problem with this is that it requires attributing humanlike capabilities for deliberate sea crossings to *Homo erectus*, not previously known to scientists as a sailor. The other, and more realistic, possibility is that anatomically modern humans from Java settled Flores Island 800,000 years ago. During their continuing life there, they made stone tools, and killed animals, including the apelike creatures now improbably called *Homo floresiensis*.

"Or it may actually be some kind of *Littlefoot*. We should keep in mind that such creatures have been observed in the Indonesian island chain."

The hobbit saga also raises problems about evolutionary theory. How can a race of people descended from *Homo erectus* be more primitive than their ancestors? Did another race of people,

191

a diminutive race coexist with Homo sapiens until the last ice age? Or is it possible that *Homo floresiensis* is descended from australopithecines which somehow arrived from Africa? These are serious questions which need to be tackled without bias in order to settle the hobbit debate.

REFERENCES:

Wikipedia entry on *Homo floresensis*

PLoS Biology- 'What is the Hobbit?' Tabitha Powlege

http://www.redorbit.com/news/science/1554324/lifes_focus_much_different_70000_years_ago/index.htm

'Bones of Contention' John Vidal, 'The Guardian' January 13, 2005

http://www.guardian.co.uk/science/2005/jan/13/research.science

PYGMY SKELETONS OF PALAU

In March 2008 more pygmy skeletons were reported from the Micronesian island of Palau which is thousands of kilometers from Flores. Lee Berger and colleagues from the University of the Witwatersrand, Rutgers University and Duke University described the fossils as being 1,400 to 3,000 years old in a study funded by the National Geographic Society Mission Programs.

At least ten burial caves have been discovered in the rock islands, with one alone yielding at least 25 individuals. They are very small with small heads, similar to Flores skeletons. One adult male would have weighed about 43 kilograms and the female around 29 kilograms.

The specimens were discovered in two caves, Ucheliungs and Omedokel which were apparently used as burial sites. Not only did Omedokel cave contain skeletons the size of *H. floresiensis,* but also larger Homo sapiens dated to between 940 and 1,080 years ago.

In the original article in PLoS One, Berger et al. wrote: "The question may be asked if the human remains from Palau represent a case of insular dwarfing in a population of *H. sapiens*, or if – as with the inferred situation on Flores Island - they may represent a separate species of small-bodied humans. While the remains from Palau share with the fossils from Flores some morphological features that are primitive for the genus *Homo* (discussed below), they also

192

possess craniofacial traits that are considered to be uniquely derived (autapomorphic) in *H. sapiens*. These features include a distinct maxillary canine fossa, a clearly delimited mandibular mental trigone (in most specimens), moderate bossing of the frontal and parietal squama, a lateral prominence on the temporal mastoid process, reduced temporal juxtamastoid eminences, and (based on a partial cranial vault preserving portions of the occipital and right and left parietals) an "*en maison*" cranial vault profile with greatest interparietal breadth high on the vault. Furthermore, at least one of the primitive features seen in some of the Palauan fossils – the distinct development of a supraorbital torus – is also seen in some modern human populations. We feel that the most parsimonious, and most reasonable, interpretation of the human fossil assemblage from Palau is that they derive from a small-bodied population of *H. sapiens* (representing either rapid insular dwarfism or a small-bodied colonizing population), and that the primitive traits they express reflect possible pliotropic or epigenetic correlates of developmental programs for small body size.

http://www.plosone.org/article/info:doi/10.1371/journal.pone.0001780

While Berger and his colleagues acknowledge some morphological similarities between the Palau and Flores skeletons, they suggest that some of the features which characterized the Flores individuals as separate species, may in fact be a common adaption in pygmy humans. However Professor Bert Roberts, co discoverer of *H. floresiensis*, from the University of Wollongong commented that the skeletons of people who died around 900 to 2900 years ago, did not indicate that island dwarfism could produce people like the hobbits.

"They have found nice, small, recently dead, modern human pygmies. I don't see how that makes any difference to the status of the *Homo floresiensis* hobbits, who don't look like modern humans," Roberts said.

The general consensus at this point in time is that the more recent Palau skeletons are of modern *H. sapiens* and bear no relationship to *H. floresiensis*. However, a recent article submitted in 'Paleontology' accuses the Palau discoverers of mismatching leg bones, teeth and brow ridges. Scientists from the University of Oregon, North Carolina University and the Australian National University refute the assumption that pygmy people once lived in Palau. Anthropologist Greg Nelson of the University of Oregon accused Berger of cowboy tactics in the way he hastily excavated the skeletons, drew conclusions and published without understanding the bigger picture.

"Our evidence indicates the earliest inhabitants of Palau were of normal stature, and it counters the evidence that Berger, et al, presented in their paper indicating there was a reduced stature population in early Palau," said Nelson. "Our research from whole bones and whole skeletons indicates that the earliest individuals in Palau were of normal stature but gracile. In other words, they were thin… I think Berger's primary mistakes were his not understanding the variation in the skeletal population in which he was working, using fragmentary remains again in a situation where he didn't understand variation, and stepping outside his own area of expertise, which, I think all scientists try not to do but sometimes we do."

http://www.scientificblogging.com/news_releases/anthropology_fight_ micronesian_dwarfs_were_not_hobbits_says_study

PART 6
GIANTS

MEGANTHROPUS
'GOLIATH'
SWARTKRANS MAN
HOLOCAUST OF AMERICAN GIANTS?
OTHER ALLEGED GIANTS FROM AROUND
THE WORLD
REX GILROY'S AUSTRALIAN GIANTS
GIANT HOAXES

MEGANTHROPUS

One of the most controversial, if not ignored, hominid fossils was discovered at Sangiran, Central Java in 1941 by von Koenigswald who was unfortunately captured by the Japanese in World War 11. However, he had managed to send a cast of the jaw fragment to Franz Weidenreich who was struck by its huge size. The jaw was about the same height as a gorilla's but much thicker, leading Weidenreich to initially classify it as a victim of acromegalic gigantism. He later rejected this theory because the jaw did not have typical pathogenic features such as an exaggerated chin and small teeth.

The jaw, *Meganthropus A/Sangiran 6,* was initially classified as *Meganthropus paleojavanicus,* by Weidenreich, who believed it was a descendant of the hominoid Gigantopithecus and eventually evolved into *Homo erectus* and then into modern Asians. This multi-regional theory of human evolution has been discarded by modern palaeoanthropology.

Another major theory, first proposed by J.T. Robinson, was that the Meganthropus jaws are South East Asian australopithecines. Although adopted by von Koenigswald and Grover Krantz, *Australopithecus paleojavanicus* has also been rejected by many palaeoanthropologists who do not believe that Australopithecus existed outside Africa. The general consensus today is that the huge jaw was related to *Homo erectus,* a tidy classification which fits in with known skeletal remains. Therefore, it is currently classified as *Homo erectus meganthropus.*

Other fossil finds, mainly from Sangiran have also been assigned to *Homo erectus meganthropus. Meganthropus B/Sangiran 8* was another jaw fragment which was around the same size and shape as the original *Me-*

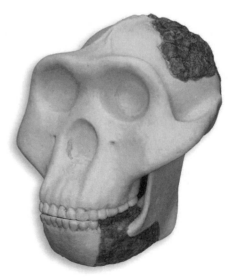

Meganthropus skull reconstructed.

ganthropus A. However, it was severely damaged, and when repaired by a Japanese/Indonesian team, was, surprisingly shown to be smaller than other known specimens of *H .erectus*. *Meganthropus C/Sangiran 33/bk 7905* was discovered in 1975 and shared some characteristics with the other mandibles.

Meganthropus D, discovered in 1993, has been dated to between 1.4 and 0.9 million years ago. Similar in shape and size to *Meganthropus A*, this mandible without teeth has been accepted as representations of the same species as *Meganthropus A* by Kranz and Sartono.

Meganthropus 11/Sangiran 31 is a skull fragment which is thicker, lower vaulted and wider than any other specimen. It had a double sagittal crest with a cranial capacity of around 800-1000 ccm.

According to Wikipedia, "Meganthropus has been the target of numerous extreme claims, none of which are supported by peer-reviewed authors. Perhaps the most common claim is that Meganthropus was a giant, one unsourced claim put them at 9 feet (2.75 m) tall and 750 to 1000 pounds (340 to 450 kilograms). No exact height has been published in a peer reviewed journal recently, and none give an indication of Meganthropus being substantially larger than *H. erectus*."

It is interesting that Wikipedia discards claims of gigantism whereas in the same article it says: "Weidenreich never made a direct size estimate of the hominid it came from, but said it was 2/3 the size of <u>*Gigantopithecus*</u>, which was twice as large as a gorilla, which would make it somewhere around 8 feet (2.44 m) tall. The jawbone was apparently used in part of <u>Grover Krantz</u>'s skull reconstruction, which was only 8.5 inches (21 centimeters) tall."

Exactly how tall does a skeleton have to be to be classified as gigantic? Also a jawbone of 8.5 inches is considerably larger than that of an adult!

'GOLIATH'

One of the tallest hominids discovered was named 'Goliath' by Lee Berger and Steve Churchill in the National Geographic program 'Searching for the Ultimate Survivor.' This program claimed that the hominid named Goliath was the largest hominid ever discovered, although other researchers claim that it is a member of the *Homo heidelbergensis* species. The specimen in question was a femur found by Berger from Hoedjiespun, South Africa and about 300,000 years old. Lee Berger reconstructed this 'giant' based on the Kabwe cranial and postcranial remains and concluded it was 6'4" tall.

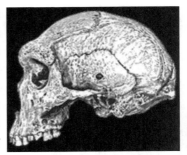

The Kabwe cranium.

The classification of *Homo heidelbergensis* is itself controversial as many researchers believe it is a European species (such as the Mauer mandible) whereas others believe it applies to all Middle Pleistocene European and African fossils a well as some from China.

Were these hominids giants? Unquestionably some individuals had a very large mass, including the Kabwe and a very broad pelvis from Sima de los Huesos in Spain. The hominids were almost two meters tall and had masses of 80 to 90 kg which makes them within the normal range of modern humans, although large for Pleistocene hominids.

The term 'giant' and 'Goliath' is rather misleading as it implies a person of huge stature well in excess of six feet. Today in industrialized countries mean body sizes are very similar to those of Middle Pleistocene humans.

SWARTKRANS MAN

The 'New York Times' announced the discovery of a giant in South Africa on December 1, 1948 by highly regarded anthropologist Dr. Robert Broom. "Discovery near Johannesburg of a new type of ape man which might have been larger than the nine-foot-tall primitive Java man found several years ago was reported today by the University of California… A local paleontologist who was working with the leader of an expedition to the area from U.C. Berkeley, Dr. Robert Broom, made the discovery. The initial discovery consisted of some extremely large dentition; two upper incisors, an upper canine and part of a lower jaw still containing three premolars as well as four molars.

"The jaw is very massive," Dr. Broom cabled. The teeth are a little larger than in the recently discovered giant man of Java (meganthropus.) The incisor and canine teeth are typically human and very unlike those of an anthropoid ape."

In his book 'Swartkrans ape-man Paranthropus crassidans' he wrote: "The jaw is really huge. The direct measurement from the condyle to the front of the chin is 165 mm. In the Heidelberg jaw the corresponding measurement is 140 mm and in a large male Bantu 135 mm. In one of the Wadjak skulls the measurement is 143 mm."

'Life' magazine ran a feature on the Swartkrans excavations on March 21, 1949. It wrote, "One was a giant heavier than Gargantua, the gorilla. The other was a pygmy no bigger than a midget. The three finds, ranking in importance with Pithecanthropus erectus and the China and Java giants help bridge the still wide gap between man and the earliest apelike creatures from which he rose."

"Greatest of the three in size and significance is the monstrous but basically human creature reconstructed above. A lumbering beast, he roamed a semidesert world inhabited by now extinct giraffes and sabre-toothed tigers. It was among their petrified bones that massive jaw fragments were found by the South African palaeontologist Robert Broom. Aided by a grant from Wendell Phillips, leader of the University of California's African Expedition, the 82 year old scientist last fall resumed his lifelong search for the remains of ancient man. The giant jaw, which he chipped from the wall of a cave in the Transvaal, was one of the greatest rewards of his four decades of exploration. Its age, he reckons, might be one million years. And small as the fragment was, its form proved that Swartkrans man was human."

'Time' magazine wrote on Dec 13, 1948, "'Wendell Phillips... described Swartkrans man as "a million dollar discovery, what we were deaming about for 14 months in Africa." The discovery of Swartkrans Man should buttress the theory, not previously accepted by all palaentologists, that nature experimented, something over a million years ago, with big, lumbering men before finally settling on the present model.'"

Initially dubbed 'Swartkrans man,' it was estimated that the hominin lived between 500,000 and five million years ago. Broom then designated fossil SK6 as *Paranthropus crassidans* or *Paranthropus robustus crassidans,*

So what became of Swartkrans man and why was he excised from the fossil record? Now the once famous Swartkrans giant is described as such: "Paranthropus stood about 1.20 meters tall and probably weighed less than 50 kilograms, and its pelvis and leg structure indicate it was bipedal," according to Professor Matt Sponheimer who also added that its brain to body size ratio is "slightly larger than the chimpanzees."

http://www.afrol.com/articles/225549

(Afrol news, 10 November)

Eventually other fossils were excavated from Swartkrans and nearby Kromdaai and *crassidans* was lumped together with other *Australopithecus robustus* specimens. Although robustus had huge molars, its incisors and brain were very small and its body was not much bigger than that of africanus. Without any other bones to support the gigantism claims, it seems that *crassidans* was destined to slide into obscurity.

HOLOCAUST OF AMERICAN GIANTS?

A thought provoking essay by Vine Deloria called 'Holocaust of the Giants' accuses the Smithsonian of destroying evidence that gigantic bones discovered in the mounds were spirited away, never to be seen again.

This section will look at such claims, concentrating mainly on 19th century reports. The next section will examine more dubious and unsupported claims of gigantism. Many of the so-called 'Giants' were tall people in excess of 6 feet, but hardly of gigantic proportions. Only those skeletal remains in excess of 7 feet (which can still be found in many humans today) will be highlighted.

OHIO
The History of Marion County, Ohio
(compiled from past accounts, published in 1883):

"Mastodonic remains are occasionally unearthed, and, from time to time, discoveries of the remains of Indian settlements are indicated by the appearance of gigantic skeletons, with the high cheek bones, powerful jaws and massive frames peculiar of the red man, who left these as the only record with which to form a clew to the history of past ages."

The History of Brown County, Ohio
(compiled from past accounts, published in 1883):

"She said also that three skeletons were found at the mouth of the Paw Paw Creek many years later, while Nim (Nimrod) Satterfield was justice of the peace. Jim Dean and some men were digging for a bridge foundation and found these bones at the lower end of the old buffalo wallow. She thought it was Dr.. Kidwell, of Fairmont, who examined them and said they were very old, perhaps thousands of years old. She said that when the skeletons were exposed to the weather for a few days, their bones turned black and began to crumble, that Squire Satterfield had them buried in the Joliffe graveyard (Rivesville). All these skeletons, she said, were measured, and found to be about eight feet long."

The Scientific American, in 1883, published the following account:

"Two miles from Mandan, on the bluffs near the junction of the Hart and Missouri Rivers, says the local newspaper, the Pioneer, is an old Cemetery of fully 100 acres in extent filled with bones of a giant race. This vast city of the dead lies just east of the Fort Lincoln road. The ground has the appearance of having been filled with trenches piled full of dead bodies, both man and beast, and covered with several feet of earth. In many places mounds from 8 to 10 feet high, and some of them 100 feet or more in length, have been thrown up and are filled with bones, broken pottery, vases of various bright colored flint, and agates ... showing the work of a people skilled in the arts and possessed of a high state of civilization. This has evidently been a grand battlefield, where thousands of men ...have fallen. ...Five miles above Mandan, on the opposite side of the Missouri, is another vast cemetery, as yet unexplored. We asked an aged Indian what his people knew of these ancient grave yards. He answered: "Me know nothing about them. They were here before the red man."

From the '*Ironton Register*', a small Ohio River town newspaper, dated 5 May 1892:

"Where Proctorville now stands was one day part of a well paved city, but I think the greatest part of it is now in the Ohio river. Only a few mounds, there; one of which was near the C. Wilgus mansion and contained a skeleton of a very large person, all double teeth, and sound, in a jaw bone that would go over the jaw with the flesh on, of a large man; The common burying ground was well filled with skeletons at a depth of about 6 feet. Part of the pavement was of boulder stone and part of well preserved brick.

"She said also that three skeletons were found at the mouth of the Paw Paw Creek many years later, while Nim (Nimrod) Satterfield was justice of the peace. Jim Dean and some men were digging for a bridge foundation and found these bones at the lower end of the old buffalo wallow. She thought it was Dr. Kidwell, of Fairmont, who examined them and said they were very old, perhaps thousands of years old. She said that when the skeletons were exposed to the weather for a few days, their bones turned black and began to crumble, that Squire Satterfield had them buried in the Joliffe graveyard (Rivesville). All these skeletons, she said, were measured, and found to be about eight feet long."

'Now and Long Ago-A History of the Marion County Area' by Glen Lough (1969)
http://www.burlingtonnews.net/ohiogiants.html

'Scientific American' August 14, 1880 Page 106:

"The Rev. Stephen Bowers notes, in the Kansas City Review of Science, the opening of an interesting mound in Brush Creek Township, Ohio. The mound was opened by the Historical Society of the township, under the immediate supervision of Dr. J.F. Everhart, of Zanesville.

"It measured sixty-four by thirty-five feet at the summit, gradually sloping in every direction and was eight feet in height. There was found in it a sort of clay coffin including the skeleton of a woman measuring eight feet in length.

Within this coffin was found also the skeleton of a child about three and a half feet in length and an image that crumbled when exposed to the atmosphere.

In another grave was found the skeleton of a man and a woman, the former measuring nine and the latter eight feet in length. In a third grave occurred two other skeletons, male and female, measuring respectively nine feet four inches and eight feet."

Seven other skeletons were found in the mound, the smallest of which measured eight feet, while others reached the enormous length of ten feet.

They were buried singly, or each in separate graves. Resting against one of the coffins was an engraved stone tablet (now in Cincinnati) from the characters on which Dr.. Everhart and Mr. Bowers are led to conclude that this giant race were sun worshippers."

'The New York Times' May 5, 1885 ran an article 'SKELETONS SEVEN FEET LONG.' It described how schoolboys opened a mound near Homer and found a skeleton. "Today further search was made, and several feet below the surface of the earth in a large vault, with stone floor and bark covering, were found four huge skeletons, three being each over seven feet in length and the other eight. The skeletons lay with their feet to the east on a bed of charcoal in which were numerous partially burned bone." The skeletons were found with stone necklaces, stone vessels and a pipe.

TENNESSEE

2ᵗʰ Annual Report of the Bureau of Ethnology to the Secretary of the Smithsonian Institution 1890-1891
(published in 1894) (explorations in the Tennessee District):

"Underneath the layer of shells the earth was very dark and appeared to be mixed with vegetable mold to the depth of 1 foot. At the bottom of this, resting on the original surface of the ground, was a very large skeleton lying horizontally at full length. Although very soft, the bones were sufficiently distinct to allow of careful measurement before attempting to remove them. The length from the base of the skull to the bones of the toes was found to be 7 feet 3 inches. It is probable, therefore, that this individual when living was fully 7½ feet high. At the head lay some small pieces of mica and a green substance, probably the oxide of copper, though no ornament or article of copper was discovered."

12ᵗʰ Annual Report of the Bureau of Ethnology to the Secretary of the Smithsonian Institution 1890-1891
(published in 1894) (explorations in Roane County, Tennessee):

"But Thomas' time was limited because of the large territory he was to explore. Under such working conditions, anomalies were put aside for future research—to be, as it has turned out, forgotten. Thomas was forced to rely on the accounts of operatives in many cases. Evidently, some of these people discerned between "Indian" burials and the burials of the Mound Builders, perhaps challenging the patience of Powell.

"No. 5, the largest of the group was carefully examined. Two feet below the surface, near the apex, was a skeleton, doubtless an intrusive Indian burial... Near the original surface, 10 or 12 feet from the center, on the lower side, lying at full length on its back, was one of the largest skeletons discovered by the Bureau agents, the length as proved by actual measurement being between 7 and 8 feet. It was clearly traceable, but crumbled to pieces immediately after removal from the hard earth in which it was encased...."

ILLINOIS
12ᵗʰ Annual Report of the Bureau of Ethnology to the Secretary of the Smithsonian Institution 1890-1891
(published in 1894) (Pike County, Illinois):

"No. 11 is now 35 by 40 feet at the base and 4 feet high. In the center, 3 feet below the surface, was a vault 8 feet long and 3 feet wide. In the bottom of this, among the decayed fragments of bark wrappings, lay a skeleton fully seven feet long, extended at full length on the back, head west. Lying in a circle above the hips were fifty-two perforated shell disks about an inch in diameter and one-eighth of an inch thick."

2ᵗʰ Annual Report of the Bureau of Ethnology to the Secretary of the Smithsonian Institution 1890-1891 (published in 1894):

"Largest in the collective series of mounds, the Great Smith Mound yielded at least two large skeletons, but at different levels of its deconstruction by Thomas' agents. It was 35 feet in height and 175 feet in diameter, and was constructed in at least two stages, according to the report. The larger of the two skeletons represented a man conceivably approaching eight feet in height when living.

"At a depth of 14 feet, a rather large human skeleton was found, which was in a partially upright position with the back against a hard clay wall...All the bones were badly decayed, except those of the left wrist, which had been preserved by two heavy copper bracelets...

"Nineteen feet from the top the bottom of this debris was reached, where, in the remains of a bark coffin, a skeleton measuring 7½ feet in length and 19 inches across the shoulders, was discovered. It lay on the bottom of the vault stretched horizontally on the back, head east, arms by the sides... Each wrist was encircled by six heavy copper bracelets...Upon the breast was a copper gorget...length, 3½ inches; greatest width 3¾ inches..."

'Historical Encyclopedia of Illinois and History of Lake County'
Edited by Newton Bateman, and Paul Selby 1902):

"Excavations... have revealed the crumbling bones of a mighty race. Samuel Miller ...is the authority for the statement that one skeleton which he assisted in unearthing was a trifle more than eight feet in length, the skull being correspondingly large, while many other skeletons measured at least seven feet.."

MISSISSIPPI
12ᵗʰ Annual Report of the Bureau of Ethnology to the Secretary of the Smithsonian Institution 1890-1891
(published in 1894) (Union County, Mississippi):

"A femur exceeding eighteen inches would indicate a man of very great height.-easily over seven feet. Femurs exceeding twenty inches have been found however. Though hindsight is said to be 20/20, Thomas' methodology was little better than a government-sanctioned dissolution of the sacred burial places. He dismantled the sanctuaries and charnel houses with the fervor of a man whose first priority was to impress his employer. From Florida to Nebraska—including twenty-three states and Canada's Manitoba region—over the next seven years he and his agents worked like men possessed of a deadline.

"A large Indian mound near the town of Gastersville, [Gastonville?—Ed.] Pa. has recently been opened and examined by a committee of scientists sent out from the Smithsonian Institute. At some depth from the surface a kind of vault was found in which was discovered the skeleton of a giant measuring seven feet two inches. His hair was coarse and jet black, and hung to the waist, the brow being ornamented with a copper crown. The skeleton was remarkably well preserved...On the stones which covered the vault were carved inscriptions, and these when deciphered, will doubtless lift the veil that now shrouds the history of the race of people that at one time inhabited this part of the American continent. The relics have been carefully packed and forwarded to the Smithsonian Institute, and they are said to be the most interesting collection ever found in the United States."

http://www.xpeditionsmagazine.com/magazine/articles/giants/holocaust.html

FURTHER REPORTS FROM NEWSPAPERS

NEW YORK
'The Daily Telegraph', Toronto, August 23, 1871:

"On Wednesday last, Rev Nathaniel Wardell, Messers. Orin Wardel and Daniel Fredenburg were digging on the farm of the latter gentleman. Which is on the banks of the Grand River, in the township of Cayuga. When they got to five or six feet below the surface, a strange sight met them. Piled in layers, one upon top of the other, some two hundred skeletons of human beings nearly perfect- around the neck of each one being a string of beads....Thee skeletons are those of men of gigantic stature, some of them measuring nine feet, very few of them being less than seven feet. Some of the thigh bones were found to be at least a foot longer than those at present known, and one of the skulls being examined completely covered the head of an ordinary person."

'Illustrations of the Ancient Monuments of Western New York'- T. Apoleon Cheney noted a twelve foot high elliptical mound around Cattaraugus County's Conewango Valley contained eight large skeletons. Most crumbled, but one femur was 28 inches long. Stone points, enamelwork and jewelry were also found. Cheney also mentioned a skeleton 7.5 feet from Chautauqua County. Inside a very old mound near Cassadaga Lake were large skeletons, one being nearly 9' tall.

PENNSYLVANIA
'AN ALLEGED FOSSIL MAN' Hazleton Penn. *Sentinel*, August 1881:

"About three miles from Ashley, a Mr. McCauley has the contract from the Wilkes-Barre Coal & Iron Co. for sinking a coal shaft. It is located near the base of the mountain and has reached a depth of 475 feet.

Saturday last, when the gang, or what is known as the second shift of men, were about retiring, after firing off a course of holes, Tom Cassidy, the foreman, descended the shaft to ascertain the result of the explosion, and was astonished to find an immense cavity in one of the sides of the shaft.

The explosion appeared to have a terrible effect and caused more damage than benefit on one side, but his astonishment was still greater increased on clearing away some of the refuse of the rock blown by the shots to discover a solid mass of rock in which appears a clearly-defined human shape of giant proportions.

All the limbs, muscles and linaments are apparent. The rock is about 16 feet in length, 18 in breadth, and about 8 in thickness. The dimensions of the human frame are giantly, measuring 12 feet in length and 4 feet across the chest.

Across the breast is the impression of a huge shield, about four feet in circumference, while the right hand clutches the broken and butt end of a large cutlass or sword.

The rock was taken out whole and is now in possession of Mr. McCauley in Ashley."

http://s8int.com/giants5.html

Binghampton, Pennsylvania:

"In the mound uncovered were found the bones of sixty-eight men which are believed to have been buried 700 years ago. The average height of these men was seven feet, while many were much taller. Further evidence of their gigantic size was found in large celts or axes hewed from stone and buried in the grave. On some of the skulls, tow inches above the perfectly formed forehead, were protuberances of bone....The skull and a few bones found in one grave were sent to the American Indian Museum."

(*New York Times*, July 14, 1916)

WEST VIRGINIA
'The New York Times' Nov 20, 1883 ran a small article entitled 'A GIANT'S REMAINS IN A MOUND'. It said "Professor Norris, the ethnologist who has been examining the mounds in this section of West Virginia for several months, the other day opened the big mound on Col. B.H. Smith's farm... The mound is 50 feet high and they dug down to the bottom... At the bottom they found the ones of a human being, measuring 7 feet in length and 19 inches across the shoulders... At the head of the chief lay another man with his hands extended before him, and bearing two bracelets of copper..."

TEXAS
The Morhiss Mound in Texas was excavated in the 1930s and 1940 and about 219 human burials were documented. The site is thoroughly discussed in this website but there is no mention of any unusual discoveries.

http://www.texasbeyondhistory.net/morhiss/evidence. html#human

But when the site was excavated in 1940, the amazing story of a giant skull was reported in the press, with pictures. The *'San Antonio Express'* published an article 'Beach Giant Skull Unearthed by WPA Workers Near Victoria.' It said, "Twice the size of the skull of a normal man, the fragments were dug up by W. Duffen, archaeologists who is excavating the mound in Victoria County under a WPA project sponsored by the University of Texas... A study is being made to determine whether the huge skull was that of a man belonging to a tribe of extraordinary men, or whether the skull was that of an abnormal member of a tribe, a case of gigantism. Several large human bones have been unearthed from the site..."

This skull was listed as 'missing' from the Texas Archaeological Research Laboratory.
http://www.sydhav.no/giants/victoria_texas.htm

MINNESOTA

On Feb. 18, 1868 an 'antideluvian giant' was uncovered in Sank Rapids. "The remains are completely petrified, and are of gigantic dimensions. The head is massive, measures thirty one and one half inches in circumference, but low in the asfrontis and very flat on top. The femur measures twenty six and a quarter inches, and the Fibuis twenty-five and a half, while the body is equally long in proportion. From the crown of the head to the sole of the foot, the length is ten feet nine and a half inches. The giant must have weighed at least 906 pounds when covered with a reasonable amount of flesh. The petrified remains, and there is nothing left but the naked bones, now weigh 304 ¼ pounds." (*NY Times*, December 25, 1868)

According to '*St Paul Pioneer Press*' May 23, 1883 ten skeletons of "Both sexes and of gigantic size" were taken from a mound at Warren, Minnesota.

'*St Paul Globe*', Aug 12, 1896 reported the skeleton of a huge man was uncovered at Beckley farm, Lake Koronis, while at Moose Island and Pine City, other giant bones were found.

WISCONSIN

A newly discovered mound was opened in Maple Creek, Wis. in 1897, "In it was found the skeleton of a man of gigantic size. The bones measured from head to foot over nine feet and were in a fair state of preservation. The skull was as large as a half bushel measure. Some finely tempered rods of copper and other relics were lying near the bones."

(*New York Times*, December 20, 1897)

NEW MEXICO

"Luiciana Quintana, on whose ranch the ancient burial plot is located, discovered two stones that bore curious inscriptions, and beneath these were found in shallow excavations the bones of a frame that could not have been less than 12 feet in length. The men who opened the grave way the forearm was 4 feet long and that in a

210

well-preserved jaw the lower teeth ranged from the size of a hickory nut to that of the largest walnut in size." (*The New York Times,* Feb 11, 1902)

UNREFERENCED REPORTS— FROM VARIOUS INTERNET SITES

These reports are largely unreferenced and may or may not be valid.

Kentucky—Two young men named White discovered a cave with large galleries. In the first gallery they found strange characters carved in the wall upon a vault which contained the remains of three skeletons measuring 8.7, 8.5 and 8.45 feet. Beside them lay three huge swords. This report is undated and the cave is supposedly on private land.

Supposed giant skeleton from a cave in Kentucky.

Ohio—Mary Sutherland, author of 'Living in the Light' has documented these giant stories from Ohio.

•In 1828 workers dug into a mound near Chesterville and found a large human skeleton. Its head was reputedly able to fit over a normal man's head with no difficulty and it had additional teeth.

•In 'Bates' mound three skeletons were found with large skulls and gigantic proportions.

•The December 17, 1891 issue of the respected journal *Nature* reported the discovery of a giant man buried 14 feet within one of Ohio's mysterious burial mounds. The enormous man's arms, jaw,

Rare photo of a giant mummified woman from Yosemite.

arms, chest and stomach were all clad in copper. Wooden antlers, also covered with copper, rested on either side of his head. His

212

mouth was filled with large pearls, and a pearl-studded necklace of bear teeth hung around his neck.

California —

•The bones of a 12-foot tall man were dug up by a group of soldiers at Lompock Rancho (near San Luis Obisbo) in 1833. The skeleton was surrounded by giant weapons and the skull had double rows of both upper and lower teeth. Unfortunately it was secretly reburied because of Indian objections.

•In 1931 skeletons from 8.5 to 10 feet long were found in the Humbolt Lake bed.

•A giant found off the Californian Coast on Santa Rosa Island in the 1800s was distinguished by double rows of teeth.

A mummified woman from Yosemite.

•The so called Yosemite mummy was found by a group of gold miners in 1885. After pulling down an artificial wall in the pursuit of treasure, they discovered two vaults which had been carved in the rock. The second vault contained a tall mummified corpse that was 6 feet 8 inches and wrapped in animal skins. Upon removing the skins, the miners were astonished to see the corpse of a lady holding a small child to her breast. The mummy was taken to be studied by scientists at Los Angeles and slipped into obscurity.

Indiana—In 1879 archaeologists dug into an ancient burial mound at Brewersville and unearthed a skeleton 9.8 feet according to the November 1975 edition of the Indianapolis News. A mica necklace hung around its neck and a human effigy of clay was standing at its feet. The Mica Giant was examined by scientists from Indiana and New York but remained in private possession until the bones, which had been stored in a grain mill, were lost in a flood in 1937.

213

•In 1925 amateurs digging at an Indian mound in Walkerton discovered eight skeletons ranging from 8 to 9 feet buried in "substantial copper armour."

Iowa—'The Kossuth Giants'

This amazing story details seven gigantic mummies found outside the Kossuth Center buried in a basalt vault. "When the light from the first torch penetrated the gloom of the ancient structure, Albert Grosslockner gasped at what he thought were seven huge and exquisitely detailed statues seated in a ring around a very large fire pit. Moving closer, he realized that the figures were not carved of stone, but were in fact the mummified remains of some giant humanoid race...The figures, were each fully ten feet tall even when measured seated in their cross legged positions. They all faced into the circle with arms folded across their legs. Upon close examination it was seen that they had double jaws of teeth in their upper and lower jaws. The foreheads were unusually low and sloping, with exceedingly prominent brows. The skin of the mysterious giants was wrinkled and tough, as though tanned, and the hair of each of them was distinctly red in color."

http://www.burlingtonnews.net/giantkossuth.html

According to the article the bodies were removed for x-ray and autopsy examination at Georg von Podegrad College which demonstrated that there was definite skeletal structure and they had been alive. Fascinating as this tale is, it is most probably a hoax as no such college exists and the town of Kossuth Center has been abandoned for many years.

Minnesota—Skeletons of "enormous dimensions" were found in mounds near Warren Minnesota.

•In Clearwater skeletons of seven giants with double teeth and receding foreheads were found in mounds.

Nevada—In 1911 red haired mummies ranging from 6.5 to 8 feet were discovered in a cave in Lovelock. In 1932 more skeletons were found in the Humboldt lake bed near Lovelock. The first was 8.5 feet

214

tall and wrapped in a gum covered fabric. The second was almost 10 feet long according to the *Lovelock Review-Miner's* article of June 19, 1931.

•A 7.7 feet skeleton was reportedly found on the Friedman ranch near Lovelock in 1939. (Reference- 'Miner' Sept 29, 1939)

•A massive femur was discovered in solid rock by prospectors near Spring Valley in 1877. The bones were embedded in hard quartzite and the bones were almost black with carbonization. Once removed from the rock, the bones proved to be a broken femur, the knee cap and joint, the lower leg bones and the complete foot bones. The large size of the bones indicated that the owner had stood over 12 feet tall. Furthermore, they were found in rock which was dated to the Jurassic era, over 185 million years old.

Alabama—According to alternative historian David Childress, three large wooden coffins were discovered in 1892 at the Crumf Cave in Alabama. These empty coffins were about 7.5 feet long, 14 to 18 inches wide and 6 to 7 inches deep. They were taken to the Smithsonian, and never seen again. It should be noted that American Indians did not bury their dead in coffins.

Death Valley—This inhospitable area straddling both California and Nevada also has its alleged gigantic remains. On August 5, 1947 the Nevada newspaper 'Hot Citizen' ran the headline 'EXPEDITION REPORTS NINE-FOOT SKELETONS'. According to the story, Howard Hill and his companions discovered a "lost civilization of men nine feet tall in California caverns" which may be the "fabled lost continent of Atlantis."

Hill tentatively dated the skeletons at 80,000 old which immediately aroused the ire of the archaeologists. Hill then claimed that the caves had been discovered in 1931 by Dr. F. Bruce Russell who literally fell in while sinking a shaft for a mine.

After the war Russell allegedly excavated the area and found the gigantic men who "apparently wore a prehistoric zoot suit- a hair garment of medium length, jacket and knee length trousers." He also described another cavern with a ritual hall "with devices and markings similar to the Masonic order."

Montana—*The Helena Independent* newspaper on October 10, 1883 published this article titled, 'Prehistoric Giant Skeleton Found.' "J. H. Hamley, a well known and reliable citizen of Barnard, Mo., writes to the Gazette, the particulars of the discovery of a GIANT skeleton, four miles southwest of that place.

A farmer named John W. Hannon, found the bones protruding from the bank of a ravine that has been cut by the action of the rains during the past years. Mr. Hannon worked several days in unearthing "the skeleton," which proved to be that of a human being whose height was twelve feet.

The head through the temples was eleven inches; from the lower part of the skull at the back to the top was fifteen inches, and the circumference forty inches. The ribs were nearly four feet long, one and three-fourths inches wide. The thigh bones were thirty-six inches long and large in proportion. When the earth was removed the ribs stood high enough to enable a man to crawl in and explore the interior of the skeleton, turn around and come out with ease.

The first joint of the greater toe above the nail, was three inches long, and the entire foot, eighteen inches in length. The skeleton lay on its face twenty feet below the surface of the ground and the toes were imbedded in the earth, indicating that the body either fell or was placed there when the ground was soft.

The left arm was passed around backward, the hand resting on the spinal column, while the right arm was stretched out to the front, and right. Some of the bones crumbled on exposure to the air, but many good specimens were preserved, and are now exhibited at, Bernard Medical school.

The skeleton is generally pronounced a relic of the prehistoric race."

Tennessee—In 1894 a mummy was found near Memphis and examined at Jackson Mound Park by three medical doctors. Drs. Williford, Turner and Pate found the body to be 9 feet ½ inches long and 400 pounds. In a 'splendid' state of preservation, its teeth were intact and the long black hair was stiff.

An earlier report in the Mobile Herald, Feb 4, 1846 described a 'huge biped' dug up in Williamson County. The skeleton was discovered 60 feet below the surface embedded in a strata of had clay. The dimensions reported were staggering: the skeleton was 19 feet high, and its thigh bones were 6 feet 6 inches. The enormous skull as 2/3 the size of a flour barrel and a coffee cup could fit within the eye sockets.

http://s8int.com/phile/giants27.html

Aleutians—The famous naturalist Ivan Sanderson once allegedly related a story about a letter he had receive from an engineer who was stationed on the Aleutian island of Shemya during World War 11. While building an airstrip many human remains were discovered. "The Alaskan mound was in fact a graveyard of gigantic human remains, consisting of crania and long leg bones. The crania measured from 22 to 24 inches from base to crown... Such a large crania would imply an immense size for a normally proportioned human. Furthermore, every skull was said to have been neatly trepanned."

(D. Childress, 'Lost Cities of North and Central America', Adventures Unlimited Press, 1992)

GIANTS FROM AROUND THE WORLD

There is no doubt that legends and myths of giants appear in nearly every culture including western modern cultures where they survive as malevolent fairy tales. Reports of gigantic bones from other countries are not as prevalent as they were in America.

The copy of a femur which was allegedly discovered in 1950s in the Euphrates Valley of Turkey near Bashan during road construction lies in the Mt Blanco Fossil Museum in Crosbyton, Texas. The femur was measured at 47.25 inches or 120 cm, which would make this giant 14-16 feet tall with 20-22 inch teeth. According to its website, this museum is "dedicated to the correct interpretation of Earth history and fossil remains, We believe that the fossil record speaks of catastrophic events happening several thousand years ago, rather than slow processes taking place over millions or billions of years as is held by the popular establishment."

http://mtblanco.com/AboutUs.htm

217

Of course, it is a museum dedicated to Young Earth Creationism which makes its exhibits suspect.

Other reports are impossible to verify and include:

- Philippines—a human skeleton of 5.2 m (17 feet) tall was unearthed at Gargayan.
- A 4 m skeleton was discovered in Sri Lanka.
- In 1936 two French archaeologists Lebeuf and Griaule found large mounds in Chad, Africa. They dug up several egg shaped funeral jars containing the remains of a giant race including their jewelry and art. These giants were called the Saos.
- "In the year 1890 some human bones of enormous size, double the ordinary in fact, were found in the tumulus of Castelnau (Herault, France) and have since been carefully examined by Prof. Kiener, who while admitting that the bones are those of a very tall race, nevertheless finds them adnormal in dimensions and apparently of morbid growth." (*New York Times*, Oct 3, 1892)
- In 1908 it was reported that a cave containing 200 skeletons of great height was found in Mexico. "Mr. Clapp arranged the bones of one of these skeletons and found the total length to be 8 feet and 11 inches. The femur reached up to his thigh, and the molars were big enough to crack a cocoanut. The head measured eighteen inches from front to back." (*NY Times*, May 4, 1908)
- Giant skeletons measuring from 10 to 12 feet were allegedly found in the State of Chihuahua, Mexico. They were reported to be in a sitting position, with the shoulders and arms bending forward and resting on the knees.
- On Feb 3, 1909 the 'New York Tribune' reported that a 'prehistoric giant of extraordinary size' had been found at Ixtapalapa, Mexico. "A peon while excavating for the foundation of a house on the estate of Ausgustin Juarez found the skeleton of a human being that is estimated to have been about 15 feet high, and who must have lived ages ago, judging from the ossified state of the bones."

REX GILROY'S AUSTRALIAN GIANTS

Rex Gilroy is an alternative Australian researcher and naturalist who has been studying anomalous phenomena for many decades. His first book 'Mysterious Australia' discusses the many zoological cryptids in this continent, as well as his belief that Australia was settled by *Homo erectus* who crossed a land bridge from Java over 150,000 years ago.

He wrote, "Besides the Aborigines (Australoids), our continent was shared by two earlier races; the proto-Australoid Wadjak Man and Solo Man. These races also shared this continent with more than one race of giant sized, stone tool-making hominids who appear to have preceded both these peoples." ('Mysterious Australia' p172)

His evidence, dismissed by archeologists, includes massive stone implements including hand axes, clubs, adzes, chisels, knives, hammer-stones and other tools ranging from 5.5 to 16.5 kilograms. Eventually Gilroy located five further 'megatool' sites across New South Wales which he names 'Early' and 'Late' phase.

Gilroy maintains that the most recent occupation layers are between 40,000 and 60,000 years old, while the 'Early' phase tools are dated from 60,000 to 180,000 B.P.

"But who were these 'Bathurst giants?' The most likely candidate would be the 'Giant Java Man' *Meganthropus palaejavanicus*, whose massive fossilized jaws and teeth, dating back 500,000 years, have been excavated in Java and China. The dimensions of these fossils suggest a race of hominids who reached heights of up to three or four meters, had immense strength and weighed up to several hundred kilograms. Many anthropologists now see Meganthropus as a giant ancestor of the later, smaller Java man, *Homo erectus*, who gradually evolved into modern man." (ibid p 173)

Gilroy either contends that the Bathurst giants reached Australia from an ancient land bridge from Java about 500,000 years ago or they were home grown. The Aborigines from the Bathurst area had a tradition about giants called the 'Bullooo' which had always inhabited the land.

Bill Gilroy recovered what he believed to be an endocast of an ancient skull near the megatool sites. The skull lacks a rear brain case and part of the lower jaw is preserved, fused to the palate. The right side of the skull is crushed. This fossil, named 'Bathurst Skull No. 1' is 25 cm in length across the dome, 19 cm in width and 18 cm in length. The skull is brachycephalic and lacks the thick brow ridges of other Java man types, leading Rex Gilroy to classify it as an offshoot of the later Solo Man.

According to Newcastle geologist Harold Webber the endocast skulls are aged between 50,000 and 100,000 years but could be even older.

In his chapter 'Stone Age Titans' Gilroy goes on to discuss the evidence for gigantic remains in excess of the Bathurst giants in Australia's prehistory. First he looked at gigantic 'footprints' from Cowra measuring almost one meter in length and 45 cm across the toes. The local Aborigines claimed it belonged to a giant of 7.6 meters who lived in Cowra during the Dreamtime. Gilroy discovered further gigantic footprints at Penrith, Bathurst, Townsville and Mt. Isa as well as trails across volcanic ash in numerous locations.

He speculated that the footprints belonged to *Gigantopithecus*, the giant apelike creature which once roamed South East Asia. "Some footprints look like giant human tracks, whereas others are more ape-like, but it is obvious that the monstrous beings who made these tracks in the sands of time stood anywhere from 4 to 6.6 or even 8.3 meters tall." (ibid p 189)

According to the Dharuk aborigines, these monsters lived and hunted throughout the Sydney area, feeding upon the megafauna which once inhabited the continent. They called the giants 'Goolagh'.

However, it is most unlikely that *Gigantopithecus* could have made any such tracks in Australia as this hominoid was only about 12 feet tall. Nor have any *Meganthropus* remains been discovered of such gigantic proportions. Despite his passion for discovery and ability to uncover anomalous curiosities, Gilroy's giants are not supported by scientific evidence.

According to Gilroy, two gigantic men encased in marble have been reportedly uncovered in both Orange and Gympie. The Orange Marble man was uncovered in a quarry in 1889 and had seven toes on each foot, one eye and no arms, although they had probably been lost in the fossilization process. Dr. C. McCarthy of Sydney declared marble man to be an actual petrified corpse, although it was later sold and shipped to Europe where it disappeared.

The second marble man, seven feet tall, was discovered by miners at Gympie, Queensland, but it had been blasted to bits. After the fractured skeleton was sent to the Brisbane museum it also disappeared forever.

It is almost impossible to locate any more details on these alleged discoveries.

REFERENCE: 'Mysterious Australia,' Rex Gilroy

GIANT HOAXES

CARDIFF GIANT

On October 16, 1869 a 10 foot 'petrified man' was uncovered from the property of William Newell in Cardiff, New York. Weighing almost 3000 pounds (1,360 kg), this giant became a local sensation which made Newell a rich man. Various scientists examined the giant in detail and insisted that he had been a living creature. Some Christian fundamentalists also defended its legitimacy because the

The Cardiff Giant.

Bible spoke of giants in the book of *Genesis*.

The Antrim Giant.

Eventually the truth emerged that the giant had been the creation of New York tobacconist George Hull, an atheist who perpetrated the hoax to embarrass fundamentalist minister Mr. Turk with whom he was arguing. Hull hired men to carve out a 10 foot long, 4.5 inch block of gypsum in Fort Dodge, Iowa, claiming it was intended for a monument to Abraham Lincoln. It was shipped to Chicago and carved by a German stonecutter who was sworn to secrecy. Various stains and acids were used to age the giant and its 'skin' was pitted with knitting needles to simulate pores.

Hull then transported the giant by rail to Newell's farm and had it buried in 1868. The whole hoax cost him $2,600. A year later Newell hired two men to ostensibly dig a well on his property and the giant once more saw the light of day.

In 1870 the giant was revealed as a fake in court and Hull confessed his hoax. Currently the Cardiff Giant rests in the Farmers' Museum in Cooperstown, New York.

ANTRIM GIANT, IRELAND

An even larger 'fossilized giant' was discovered by miner Mr. Dyer in County Antrim, Ireland in 1895 when prospecting for iron ore. Its measurements were: length 12'2", chest girth 6'6" and length of arms 4'6". There are six toes on the right foot and the giant weighs over two tons.

Dyer showed his giant in Dublin, Liverpool and Manchester, attracting local and scientific interest. Unfortunately no scientist ever examined the giant as it was 'lost' soon after this photo was taken. Its current whereabouts are unknown although creationists still use the Irish Giant as convincing evidence of the literal truth of Genesis.

PHOTOSHOP HOAXES

In 2002 a Photoshop competition was launched with these instructions: "You are to create an archaeological hoax. Your job is to show a picture of an archaeological discovery that looks so real,

had it not appeared at Worth1000, people might have done a double take." Quite a few giant photos were produced and in no time at all went viral as evidence of giant skeletons. The following images are all deliberate Photoshop hoaxes used in this competition. http://www.ancient-wisdom.co.uk/giants.htm#giantsinarchaeology

A Photoshop giant hoax photo.

Around 2000 an Internet story about a giant skeleton which had been allegedly found in Saudi Arabia by ARAMCO oil exploration company with an accompanying photo was quickly exposed as a hoax. Stories soon surfaced that the story of the discovery of a giant skeleton was not a hoax although the accompanying photo was. According to investigator Dr. Richard Pauley of Fellowship University, ARAMOC did uncover a large skeleton in southeast Saudi Arabia in 2000 but the religious police, aided by the geological branch of the oil industry, covered it up and confiscated all photos. One unnamed

One of many Photoshop giant hoax photos.

223

source managed to surreptitiously copy one of the photos taken at the site from a digicam to his laptop before the police confiscated all cameras.

Pauley said, "As you can see, the skeleton in the photo is of a more reasonable size—I estimate he or she is between 15 and 20 feet tall- and in line with what we would expect from other research. The weathering on it is consistent with the technician's description that the skull was originally found partially uncovered by the winds. And the look of horror on its face

A Photoshop giant hoax photo.

is consistent with the realization of sudden burial by some major disaster."

http://www.evcforum.net/cgi-bin/dm.cgi?action=msg&f=25&t=1591&m=1

 http://www.sydhav.no/giants/saudi_arabia.htm

REFERENCES:

Vine Deloria, 'Holocaust of the Giants'

http://www.xpeditionsmagazine.com/magazine/articles/giants/holocaust.html

A Photoshop giant hoax photo.

http://www.texasbeyondhistory.net/morhiss/evidence.html#human
'Mysterious Australia,' Rex Gilroy, URU Publications, 2000
johnhawks.net/weblog/fossils/middle/body_mass_2005.html
D. Childress, 'Lost Cities of North and Central America', Adventures
Unlimited Press, 1992
http://www.burlingtonnews.net/giantkossuth.html

One of many Photoshop giant hoax photos.

A Photoshop giant hoax photo.

GLOSSARY

Australopithecus—an extinct genus of hominids which lived in Africa from about 4 to 2 million years ago. Basically there were two distinct species- gracile and robust.

Creationism—Belief in the literal interpretation of the Biblical account of the creation of the universe and all living things. Subgroups include New Earth, Old Earth and Christian Science.

Darwinism—A theory of biological evolution developed by Charles Darwin and others, stating that all species of organisms arise and develop through the natural selection of small, inherited variations that increase the individual's ability to compete, survive, and reproduce. Also called *Darwinian theory.*

Dendrochronology —The study of climate changes and past events by comparing the successive annual growth rings of trees or old timber.

Dolichocephaloids—long headed skulls as opposed to brachycephalic skulls which are broader.

Gigantopithecus—an extinct giant ape from China and Vietnam.

Heidelberg Man —an extinct species of the genus Homo found in Germany.

Hominid—A primate of the family Hominidae, of which *Homo sapiens* is the only extant species.

Hominin—the group consisting of modern humans, extinct human species and all our immediate ancestors (including members of the genera *Homo, Australopithecus, Paranthropus* and *Ardipithecus*).

Homo antecessor —an extinct Homo species dating from 1.2 million to 800,000 years ago in Spain.

Homo erectus—an extinct species which originated in Africa about 1.8 million years ago and spread to Europe and Asia.

Homo floresiensis —the so called hobbits of Flores, Indonesia.

Homo georgicus—a species of the genus Homo found in Georgia about 1.8 million years old.

Homo habilis—is a species of the genus Homo, which lived from approximately 2.3 to 1.4 million years ago in Africa.

Homo sapiens idaltu—the oldest Homo sapiens who lived about 160,000 years ago in Africa.

Intelligent Design—A theory that universal and biological forces result from purposeful design by an intelligent being rather than random forces. It is an offshoot of Creationism.

Intervention Theory—A belief that life was deliberately introduced to this planet by extraterrestrials and espoused by Sitchin, von Daniken and Pye.

Java man—*Homo erectus* found in Java, Indonesia.

Kennewick man—is a 9,300 year old American human with non Amerindian features.

Linnanean taxomony—a system of classification based upon Carolus Linnaeus.

Macroevolution—large scale evolution which occurs over long geological periods and results in the formation of new taxonomic groups or species.

Meganthropus—An extinct hominin of large proportions from Indonesia.

Microevolution—Evolutionary change below the level of the species, resulting from relatively small genetic variations.

Multiregional hypothesis—This hypothesis argues that our earliest hominid ancestors radiated out from Africa and *Homo sapiens* evolved from several different groups of *Homo erectus* in several places throughout the world.

Mungo man—the oldest skeletal remains found in Australia dated to about 45,000 years.

Neanderthal—an extinct Homo species which lived in Europe, the Middle East and Asia.

Nutcracker Man—fossilized cranium of species *Paranthropus boisei.*

Out of Africa theory—The hypothesis that every human is descended from a small group in Africa.

Peking Man—*Homo erectus* found in China.

Starchild—an anomalous skull in the possession of Lloyd Pye which may have non human DNA.

Stratigraphy—the study of rock strata, including the distribution, deposition and age of sedimentary rocks.

Xinjiang mummies—a group of mummies with Caucasian features uncovered in western China.

Author Karen Mutton with Lloyd Pye and the Starchild skull, Sydney, Australia.

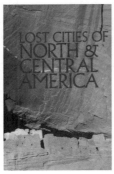

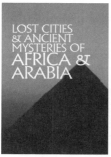

LOST CITIES & ANCIENT MYSTERIES OF AFRICA & ARABIA
by David Hatcher Childress
Childress continues his world-wide quest for lost cities and ancient mysteries. Join him as he discovers forbidden cities in the Empty Quarter of Arabia; "Atlantean" ruins in Egypt and the Kalahari desert; a mysterious, ancient empire in the Sahara; and more. This is the tale of an extraordinary life on the road: across war-torn countries, Childress searches for King Solomon's Mines, living dinosaurs, the Ark of the Covenant and the solutions to some of the fantastic mysteries of the past.
423 PAGES. 6x9 PAPERBACK. ILLUSTRATED. $14.95. CODE: AFA

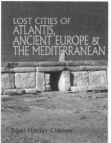

LOST CITIES OF ATLANTIS, ANCIENT EUROPE & THE MEDITERRANEAN
by David Hatcher Childress
Childress takes the reader in search of sunken cities in the Mediterranean; across the Atlas Mountains in search of Atlantean ruins; to remote islands in search of megalithic ruins; to meet living legends and secret societies. From Ireland to Turkey, Morocco to Eastern Europe, and around the remote islands of the Mediterranean and Atlantic, Childress takes the reader on an astonishing quest for mankind's past. Ancient technology, cataclysms, megalithic construction, lost civilizations and devastating wars of the past are all explored in this book.
524 PAGES. 6x9 PAPERBACK. ILLUSTRATED. $16.95. CODE: MED

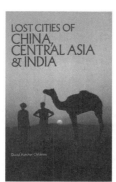

LOST CITIES OF CHINA, CENTRAL ASIA & INDIA
by David Hatcher Childress
Like a real life "Indiana Jones," maverick archaeologist David Childress takes the reader on an incredible adventure across some of the world's oldest and most remote countries in search of lost cities and ancient mysteries. Discover ancient cities in the Gobi Desert; hear fantastic tales of lost continents, vanished civilizations and secret societies bent on ruling the world; visit forgotten monasteries in forbidding snow-capped mountains with strange tunnels to mysterious subterranean cities! A unique combination of far-out exploration and practical travel advice, it will astound and delight the experienced traveler or the armchair voyager.
429 PAGES. 6x9 PAPERBACK. ILLUSTRATED. FOOTNOTES & BIBLIOGRAPHY. $14.95. CODE: CHI

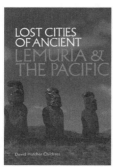

LOST CITIES OF ANCIENT LEMURIA & THE PACIFIC
by David Hatcher Childress
Was there once a continent in the Pacific? Called Lemuria or Pacifica by geologists, Mu or Pan by the mystics, there is now ample mythological, geological and archaeological evidence to "prove" that an advanced and ancient civilization once lived in the central Pacific. Maverick archaeologist and explorer David Hatcher Childress combs the Indian Ocean, Australia and the Pacific in search of the surprising truth about mankind's past. Contains photos of the underwater city on Pohnpei; explanations on how the statues were levitated around Easter Island in a clockwise vortex movement; tales of disappearing islands; Egyptians in Australia; and more.
379 PAGES. 6x9 PAPERBACK. ILLUSTRATED. FOOTNOTES & BIBLIOGRAPHY. $14.95. CODE: LEM

SUNKEN REALMS
A Survey of Underwater Ruins Around the World
By Karen Mutton

Australian researcher Karen Mutton begins by discussing some of the causes for sunken ruins: super-floods; volcanoes; earthquakes at the end of the last great flood; plate tectonics and other theories. From there she launches into a worldwide cataloging of underwater ruins by region. She begins with the many underwater cities in the Mediterranean, and then moves into northern Europe and the North Atlantic. Places covered in this book include: Tartessos; Cadiz; Morocco; Alexandria; The Bay of Naples; Libya; Phoenician and Egyptian sites; Roman era sites; Yarmuta, Lebanon; Cyprus; Malta; Thule & Hyperborea; Canary and Azore Islands; Bahamas; Cuba; Bermuda; Peru; Micronesia; Japan; Indian Ocean; Sri Lanka Land Bridge; Lake Titicaca; and inland lakes in Scotland, Russia, Iran, China, Wisconsin, Florida and more.

320 Pages. 6x9 Paperback. Illustrated. Bibliography. $20.00. Code: SRLM

TECHNOLOGY OF THE GODS
The Incredible Sciences of the Ancients
by David Hatcher Childress

Childress looks at the technology that was allegedly used in Atlantis and the theory that the Great Pyramid of Egypt was originally a gigantic power station. He examines tales of ancient flight and the technology that it involved; how the ancients used electricity; megalithic building techniques; the use of crystal lenses and the fire from the gods; evidence of various high tech weapons in the past, including atomic weapons; ancient metallurgy and heavy machinery; the role of modern inventors such as Nikola Tesla in bringing ancient technology back into modern use; impossible artifacts; and more.

356 PAGES. 6x9 PAPERBACK. ILLUSTRATED. BIBLIOGRAPHY. $16.95. CODE: TGOD

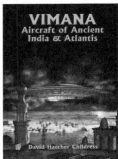

VIMANA AIRCRAFT OF ANCIENT INDIA & ATLANTIS
by David Hatcher Childress, introduction by Ivan T. Sanderson

In this incredible volume on ancient India, authentic Indian texts such as the *Ramayana* and the *Mahabharata* are used to prove that ancient aircraft were in use more than four thousand years ago. Included in this book is the entire Fourth Century BC manuscript *Vimaanika Shastra* by the ancient author Maharishi Bharadwaaja. Also included are chapters on Atlantean technology, the incredible Rama Empire of India and the devastating wars that destroyed it.

334 PAGES. 6x9 PAPERBACK. ILLUSTRATED. $15.95. CODE: VAA

LOST CONTINENTS & THE HOLLOW EARTH
I Remember Lemuria and the Shaver Mystery
by David Hatcher Childress & Richard Shaver

Shaver's rare 1948 book *I Remember Lemuria* is reprinted in its entirety, and the book is packed with illustrations from Ray Palmer's *Amazing Stories* magazine of the 1940s. Palmer and Shaver told of tunnels running through the earth—tunnels inhabited by the Deros and Teros, humanoids from an ancient spacefaring race that had inhabited the earth, eventually going underground, hundreds of thousands of years ago. Childress discusses the famous hollow earth books and delves deep into whatever reality may be behind the stories of tunnels in the earth. Operation High Jump to Antarctica in 1947 and Admiral Byrd's bizarre statements, tunnel systems in South America and Tibet, the underground world of Agartha, the belief of UFOs coming from the South Pole, more.

344 PAGES. 6x9 PAPERBACK. ILLUSTRATED. $16.95. CODE: LCHE

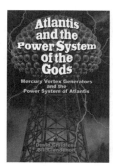

ATLANTIS & THE POWER SYSTEM OF THE GODS
by David Hatcher Childress and Bill Clendenon
Childress' fascinating analysis of Nikola Tesla's broadcast system in light of Edgar Cayce's "Terrible Crystal" and the obelisks of ancient Egypt and Ethiopia. Includes: Atlantis and its crystal power towers that broadcast energy; how these incredible power stations may still exist today; inventor Nikola Tesla's nearly identical system of power transmission; Mercury Proton Gyros and mercury vortex propulsion; more. Richly illustrated, and packed with evidence that Atlantis not only existed—it had a world-wide energy system more sophisticated than ours today.
246 PAGES. 6x9 PAPERBACK. ILLUSTRATED. $15.95. CODE: APSG

THE ANTI-GRAVITY HANDBOOK
edited by David Hatcher Childress

The new expanded compilation of material on Anti-Gravity, Free Energy, Flying Saucer Propulsion, UFOs, Suppressed Technology, NASA Cover-ups and more. Highly illustrated with patents, technical illustrations and photos. This revised and expanded edition has more material, including photos of Area 51, Nevada, the government's secret testing facility. This classic on weird science is back in a new format!
230 PAGES. 7x10 PAPERBACK. ILLUSTRATED. $16.95. CODE: AGH

ANTI–GRAVITY & THE WORLD GRID
Is the earth surrounded by an intricate electromagnetic grid network offering free energy? This compilation of material on ley lines and world power points contains chapters on the geography, mathematics, and light harmonics of the earth grid. Learn the purpose of ley lines and ancient megalithic structures located on the grid. Discover how the grid made the Philadelphia Experiment possible. Explore the Coral Castle and many other mysteries, including acoustic levitation, Tesla Shields and scalar wave weaponry. Browse through the section on anti-gravity patents, and research resources.
274 PAGES. 7x10 PAPERBACK. ILLUSTRATED. $14.95. CODE: AGW

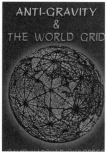

ANTI–GRAVITY & THE UNIFIED FIELD
edited by David Hatcher Childress
Is Einstein's Unified Field Theory the answer to all of our energy problems? Explored in this compilation of material is how gravity, electricity and magnetism manifest from a unified field around us. Why artificial gravity is possible; secrets of UFO propulsion; free energy; Nikola Tesla and anti-gravity airships of the 20s and 30s; flying saucers as superconducting whirls of plasma; anti-mass generators; vortex propulsion; suppressed technology; government cover-ups; gravitational pulse drive; spacecraft & more.
240 PAGES. 7x10 PAPERBACK. ILLUSTRATED. $14.95. CODE: AGU

THE TIME TRAVEL HANDBOOK
A Manual of Practical Teleportation & Time Travel
edited by David Hatcher Childress
The Time Travel Handbook takes the reader beyond the government experiments and deep into the uncharted territory of early time travellers such as Nikola Tesla and Guglielmo Marconi and their alleged time travel experiments, as well as the Wilson Brothers of EMI and their connection to the Philadelphia Experiment—the U.S. Navy's forays into invisibility, time travel, and teleportation. Childress looks into the claims of time travelling individuals, and investigates the unusual claim that the pyramids on Mars were built in the future and sent back in time. A highly visual, large format book, with patents, photos and schematics. Be the first on your block to build your own time travel device!
316 PAGES. 7x10 PAPERBACK. ILLUSTRATED. $16.95. CODE: TTH

MAPS OF THE ANCIENT SEA KINGS
Evidence of Advanced Civilization in the Ice Age
by Charles H. Hapgood
Charles Hapgood has found the evidence in the Piri Reis Map that shows Antarctica, the Hadji Ahmed map, the Oronteus Finaeus and other amazing maps. Hapgood concluded that these maps were made from more ancient maps from the various ancient archives around the world, now lost. Not only were these unknown people more advanced in mapmaking than any people prior to the 18th century, it appears they mapped all the continents. The Americas were mapped thousands of years before Columbus. Antarctica was mapped when its coasts were free of ice!
316 PAGES. 7x10 PAPERBACK. ILLUSTRATED. BIBLIOGRAPHY & INDEX. $19.95. CODE: MASK

PATH OF THE POLE
Cataclysmic Pole Shift Geology
by Charles H. Hapgood
Maps of the Ancient Sea Kings author Hapgood's classic book *Path of the Pole* is back in print! Hapgood researched Antarctica, ancient maps and the geological record to conclude that the Earth's crust has slipped on the inner core many times in the past, changing the position of the pole. *Path of the Pole* discusses the various "pole shifts" in Earth's past, giving evidence for each one, and moves on to possible future pole shifts.
356 PAGES. 6x9 PAPERBACK. ILLUSTRATED. $16.95. CODE: POP

SECRETS OF THE HOLY LANCE
The Spear of Destiny in History & Legend
by Jerry E. Smith
Secrets of the Holy Lance traces the Spear from its possession by Constantine, Rome's first Christian Caesar, to Charlemagne's claim that with it he ruled the Holy Roman Empire by Divine Right, and on through two thousand years of kings and emperors, until it came within Hitler's grasp—and beyond! Did it rest for a while in Antarctic ice? Is it now hidden in Europe, awaiting the next person to claim its awesome power? Neither debunking nor worshiping, *Secrets of the Holy Lance* seeks to pierce the veil of myth and mystery around the Spear. Mere belief that it was infused with magic by virtue of its shedding the Savior's blood has made men kings. But what if it's more? What are "the powers it serves"?
312 PAGES. 6x9 PAPERBACK. ILLUSTRATED. BIBLIOGRAPHY. $16.95. CODE: SOHL

THE FANTASTIC INVENTIONS OF NIKOLA TESLA
by Nikola Tesla with additional material by
David Hatcher Childress
This book is a readable compendium of patents, diagrams, photos and explanations of the many incredible inventions of the originator of the modern era of electrification. In Tesla's own words are such topics as wireless transmission of power, death rays, and radio-controlled airships. In addition, rare material on a secret city built at a remote jungle site in South America by one of Tesla's students, Guglielmo Marconi. Marconi's secret group claims to have built flying saucers in the 1940s and to have gone to Mars in the early 1950s! Incredible photos of these Tesla craft are included. •His plan to transmit free electricity into the atmosphere. •How electrical devices would work using only small antennas. •Why unlimited power could be utilized anywhere on earth. •How radio and radar technology can be used as death-ray weapons in Star Wars.
342 PAGES. 6x9 PAPERBACK. ILLUSTRATED. $16.95. CODE: FINT

REICH OF THE BLACK SUN
Nazi Secret Weapons & the Cold War Allied Legend
by Joseph P. Farrell

Why were the Allies worried about an atom bomb attack by the Germans in 1944? Why did the Soviets threaten to use poison gas against the Germans? Why did Hitler in 1945 insist that holding Prague could win the war for the Third Reich? Why did US General George Patton's Third Army race for the Skoda works at Pilsen in Czechoslovakia instead of Berlin? Why did the US Army not test the uranium atom bomb it dropped on Hiroshima? Why did the Luftwaffe fly a non-stop round trip mission to within twenty miles of New York City in 1944? *Reich of the Black Sun* takes the reader on a scientific-historical journey in order to answer these questions. Arguing that Nazi Germany actually won the race for the atom bomb in late 1944,

352 PAGES. 6x9 PAPERBACK. ILLUSTRATED. BIBLIOGRAPHY. $16.95. CODE: ROBS

THE GIZA DEATH STAR
The Paleophysics of the Great Pyramid & the Military Complex at Giza
by Joseph P. Farrell

Was the Giza complex part of a military installation over 10,000 years ago? Chapters include: An Archaeology of Mass Destruction, Thoth and Theories; The Machine Hypothesis; Pythagoras, Plato, Planck, and the Pyramid; The Weapon Hypothesis; Encoded Harmonics of the Planck Units in the Great Pyramid; High Freqquency Direct Current "Impulse" Technology; The Grand Gallery and its Crystals: Gravito-acoustic Resonators; The Other Two Large Pyramids; the "Causeways," and the "Temples"; A Phase Conjugate Howitzer; Evidence of the Use of Weapons of Mass Destruction in Ancient Times; more.

290 PAGES. 6x9 PAPERBACK. ILLUSTRATED. $16.95. CODE: GDS

THE GIZA DEATH STAR DEPLOYED
The Physics & Engineering of the Great Pyramid
by Joseph P. Farrell

Farrell expands on his thesis that the Great Pyramid was a maser, designed as a weapon and eventually deployed—with disastrous results to the solar system. Includes: Exploding Planets: A Brief History of the Exoteric and Esoteric Investigations of the Great Pyramid; No Machines, Please!; The Stargate Conspiracy; The Scalar Weapons; Message or Machine?; A Tesla Analysis of the Putative Physics and Engineering of the Giza Death Star; Cohering the Zero Point, Vacuum Energy, Flux: Feedback Loops and Tetrahedral Physics; and more.

290 PAGES. 6x9 PAPERBACK. ILLUSTRATED. $16.95. CODE: GDSD

THE GIZA DEATH STAR DESTROYED
The Ancient War For Future Science
by Joseph P. Farrell

Farrell moves on to events of the final days of the Giza Death Star and its awesome power. These final events, eventually leading up to the destruction of this giant machine, are dissected one by one, leading us to the eventual abandonment of the Giza Military Complex—an event that hurled civilization back into the Stone Age. Chapters include: The Mars-Earth Connection; The Lost "Root Races" and the Moral Reasons for the Flood; The Destruction of Krypton: The Electrodynamic Solar System, Exploding Planets and Ancient Wars; Turning the Stream of the Flood: the Origin of Secret Societies and Esoteric Traditions; The Quest to Recover Ancient Mega-Technology; Non-Equilibrium Paleophysics; Monatomic Paleophysics; Frequencies, Vortices and Mass Particles; "Acoustic" Intensity of Fields; The Pyramid of Crystals; tons more.

292 pages. 6x9 paperback. Illustrated. $16.95. Code: GDES

THE TESLA PAPERS
Nikola Tesla on Free Energy & Wireless Transmission of Power
by Nikola Tesla, edited by David Hatcher Childress
David Hatcher Childress takes us into the incredible world of Nikola Tesla and his amazing inventions. Tesla's fantastic vision of the future, including wireless power, anti-gravity, free energy and highly advanced solar power. Also included are some of the papers, patents and material collected on Tesla at the Colorado Springs Tesla Symposiums, including papers on: •The Secret History of Wireless Transmission •Tesla and the Magnifying Transmitter •Design and Construction of a Half-Wave Tesla Coil •Electrostatics: A Key to Free Energy •Progress in Zero-Point Energy Research •Electromagnetic Energy from Antennas to Atoms •Tesla's Particle Beam Technology •Fundamental Excitatory Modes of the Earth-Ionosphere Cavity
325 PAGES. 8x10 PAPERBACK. ILLUSTRATED. $16.95. CODE: TTP

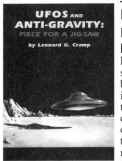

UFOS AND ANTI-GRAVITY
Piece For A Jig-Saw
by Leonard G. Cramp
Leonard G. Cramp's 1966 classic book on flying saucer propulsion and suppressed technology is a highly technical look at the UFO phenomena by a trained scientist. Cramp first introduces the idea of 'anti-gravity' and introduces us to the various theories of gravitation. He then examines the technology necessary to build a flying saucer and examines in great detail the technical aspects of such a craft. Cramp's book is a wealth of material and diagrams on flying saucers, anti-gravity, suppressed technology, G-fields and UFOs. Chapters include Crossroads of Aerodymanics, Aerodynamic Saucers, Limitations of Rocketry, Gravitation and the Ether, Gravitational Spaceships, G-Field Lift Effects, The Bi-Field Theory, VTOL and Hovercraft, Analysis of UFO photos, more.
388 PAGES. 6x9 PAPERBACK. ILLUSTRATED. $16.95. CODE: UAG

THE COSMIC MATRIX
Piece for a Jig-Saw, Part Two
by Leonard G. Cramp
Cramp examines anti-gravity effects and theorizes that this super-science used by the craft—described in detail in the book—can lift mankind into a new level of technology, transportation and understanding of the universe. The book takes a close look at gravity control, time travel, and the interlocking web of energy between all planets in our solar system with Leonard's unique technical diagrams. A fantastic voyage into the present and future!
364 PAGES. 6x9 PAPERBACK. ILLUSTRATED. BIBLIOGRAPHY. $16.00. CODE: CMX

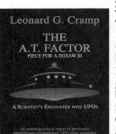

THE A.T. FACTOR
A Scientists Encounter with UFOs
by Leonard Cramp
British aerospace engineer Cramp began much of the scientific anti-gravity and UFO propulsion analysis back in 1955 with his landmark book *Space, Gravity & the Flying Saucer* (out-of-print and rare). In this final book, Cramp brings to a close his detailed and controversial study of UFOs and Anti-Gravity.
324 PAGES. 6x9 PAPERBACK. ILLUSTRATED. BIBLIOGRAPHY. INDEX. $16.95. CODE: ATF

THE FREE-ENERGY DEVICE HANDBOOK
A Compilation of Patents and Reports
by David Hatcher Childress

A large-format compilation of various patents, papers, descriptions and diagrams concerning free-energy devices and systems. *The Free-Energy Device Handbook* is a visual tool for experimenters and researchers into magnetic motors and other "over-unity" devices. With chapters on the Adams Motor, the Hans Coler Generator, cold fusion, superconductors, "N" machines, space-energy generators, Nikola Tesla, T. Townsend Brown, and the latest in free-energy devices. Packed with photos, technical diagrams, patents and fascinating information, this book belongs on every science shelf.

292 PAGES. 8x10 PAPERBACK. ILLUSTRATED. $16.95. CODE: FEH

THE ENERGY GRID
Harmonic 695, The Pulse of the Universe
by Captain Bruce Cathie

This is the breakthrough book that explores the incredible potential of the Energy Grid and the Earth's Unified Field all around us. Cathie's first book, *Harmonic 33*, was published in 1968 when he was a commercial pilot in New Zealand. Since then, Captain Bruce Cathie has been the premier investigator into the amazing potential of the infinite energy that surrounds our planet every microsecond. Cathie investigates the Harmonics of Light and how the Energy Grid is created. In this amazing book are chapters on UFO Propulsion, Nikola Tesla, Unified Equations, the Mysterious Aerials, Pythagoras & the Grid, Nuclear Detonation and the Grid, Maps of the Ancients, an Australian Stonehenge examined, more.

255 PAGES. 6x9 TRADEPAPER. ILLUSTRATED. $15.95. CODE: TEG

THE BRIDGE TO INFINITY
Harmonic 371244
by Captain Bruce Cathie

Cathie has popularized the concept that the earth is crisscrossed by an electromagnetic grid system that can be used for anti-gravity, free energy, levitation and more. The book includes a new analysis of the harmonic nature of reality, acoustic levitation, pyramid power, harmonic receiver towers and UFO propulsion. It concludes that today's scientists have at their command a fantastic store of knowledge with which to advance the welfare of the human race.

204 PAGES. 6x9 TRADEPAPER. ILLUSTRATED. $14.95. CODE: BTF

THE HARMONIC CONQUEST OF SPACE
by Captain Bruce Cathie

Chapters include: Mathematics of the World Grid; the Harmonics of Hiroshima and Nagasaki; Harmonic Transmission and Receiving; the Link Between Human Brain Waves; the Cavity Resonance between the Earth; the Ionosphere and Gravity; Edgar Cayce—the Harmonics of the Subconscious; Stonehenge; the Harmonics of the Moon; the Pyramids of Mars; Nikola Tesla's Electric Car; the Robert Adams Pulsed Electric Motor Generator; Harmonic Clues to the Unified Field; and more. Also included are tables showing the harmonic relations between the earth's magnetic field, the speed of light, and anti-gravity/gravity acceleration at different points on the earth's surface. New chapters in this edition on the giant stone spheres of Costa Rica, Atomic Tests and Volcanic Activity, and a chapter on Ayers Rock analysed with Stone Mountain, Georgia.

248 PAGES. 6x9. PAPERBACK. ILLUSTRATED. BIBLIOGRAPHY. $16.95. CODE: HCS

GRAVITATIONAL MANIPULATION OF DOMED CRAFT
UFO Propulsion Dynamics
by Paul E. Potter

Potter's precise and lavish illustrations allow the reader to enter directly into the realm of the advanced technological engineer and to understand, quite straightforwardly, the aliens' methods of energy manipulation: their methods of electrical power generation; how they purposely designed their craft to employ the kinds of energy dynamics that are exclusive to space (discoverable in our astrophysics) in order that their craft may generate both attractive and repulsive gravitational forces; their control over the mass-density matrix surrounding their craft enabling them to alter their physical dimensions and even manufacture their own frame of reference in respect to time. Includes a 16-page color insert.

624 pages. 7x10 Paperback. Illustrated. References. $24.00. Code: GMDC

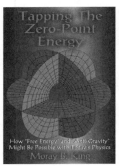

TAPPING THE ZERO POINT ENERGY
Free Energy & Anti-Gravity in Today's Physics
by Moray B. King

King explains how free energy and anti-gravity are possible. The theories of the zero point energy maintain there are tremendous fluctuations of electrical field energy imbedded within the fabric of space. This book tells how, in the 1930s, inventor T. Henry Moray could produce a fifty kilowatt "free energy" machine; how an electrified plasma vortex creates anti-gravity; how the Pons/Fleischmann "cold fusion" experiment could produce tremendous heat without fusion; and how certain experiments might produce a gravitational anomaly.

180 PAGES. 5x8 PAPERBACK. ILLUSTRATED. $12.95. CODE: TAP

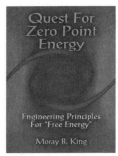

QUEST FOR ZERO-POINT ENERGY
Engineering Principles for "Free Energy"
by Moray B. King

King expands, with diagrams, on how free energy and anti-gravity are possible. The theories of zero point energy maintain there are tremendous fluctuations of electrical field energy embedded within the fabric of space. King explains the following topics: TFundamentals of a Zero-Point Energy Technology; Vacuum Energy Vortices; The Super Tube; Charge Clusters: The Basis of Zero-Point Energy Inventions; Vortex Filaments, Torsion Fields and the Zero-Point Energy; Transforming the Planet with a Zero-Point Energy Experiment; Dual Vortex Forms: The Key to a Large Zero-Point Energy Coherence. Packed with diagrams, patents and photos.

224 PAGES. 6x9 PAPERBACK. ILLUSTRATED. $14.95. CODE: QZPE

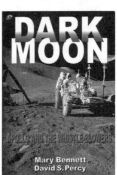

DARK MOON
Apollo and the Whistleblowers
by Mary Bennett and David Percy

Did you know a second craft was going to the Moon at the same time as Apollo 11? Do you know that potentially lethal radiation is prevalent throughout deep space? Do you know there are serious discrepancies in the account of the Apollo 13 'accident'? Did you know that 'live' color TV from the Moon was not actually live at all? Did you know that the Lunar Surface Camera had no viewfinder? Do you know that lighting was used in the Apollo photographs—yet no lighting equipment was taken to the Moon? All these questions, and more, are discussed in great detail by British researchers Bennett and Percy in *Dark Moon*, the definitive book (nearly 600 pages) on the possible faking of the Apollo Moon missions. Tons of NASA photos analyzed for possible deceptions.

568 PAGES. 6x9 PAPERBACK. ILLUSTRATED. BIBLIOGRAPHY. INDEX. $32.00. CODE: DMO

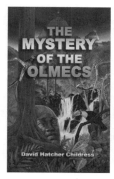

THE MYSTERY OF THE OLMECS
by David Hatcher Childress

The Olmecs were not acknowledged to have existed as a civilization until an international archeological meeting in Mexico City in 1942. Now, the Olmecs are slowly being recognized as the Mother Culture of Mesoamerica, having invented writing, the ball game and the "Mayan" Calendar. But who were the Olmecs? Where did they come from? What happened to them? How sophisticated was their culture? Why are many Olmec statues and figurines seemingly of foreign peoples such as Africans, Europeans and Chinese? Is there a link with Atlantis? In this heavily illustrated book, join Childress in search of the lost cities of the Olmecs! Chapters include: The Mystery of Quizuo; The Mystery of Transoceanic Trade; The Mystery of Cranial Deformation; more.

296 PAGES. 6x9 PAPERBACK. ILLUSTRATED. BIBLIOGRAPHY. COLOR SECTION. $20.00. CODE: MOLM

THE LAND OF OSIRIS
An Introduction to Khemitology
by Stephen S. Mehler

Was there an advanced prehistoric civilization in ancient Egypt who built the great pyramids and carved the Great Sphinx? Did the pyramids serve as energy devices and not as tombs for kings? Mehler has uncovered an indigenous oral tradition that still exists in Egypt, and has been fortunate to have studied with a living master of this tradition, Abd'El Hakim Awyan. Mehler has also been given permission to present these teachings to the Western world, teachings that unfold a whole new understanding of ancient Egypt . Chapters include: Egyptology and Its Paradigms; Asgat Nefer—The Harmony of Water; Khemit and the Myth of Atlantis; The Extraterrestrial Question; more.

272 PAGES. 6x9 PAPERBACK. ILLUSTRATED. COLOR SECTION. BIBLIOGRAPHY. $18.00 CODE: LOOS

ABOMINABLE SNOWMEN: LEGEND COME TO LIFE
The Story of Sub-Humans on Six Continents from the Early Ice Age Until Today
by Ivan T. Sanderson

Do "Abominable Snowmen" exist? Prepare yourself for a shock. In the opinion of one of the world's leading naturalists, not one, but possibly four kinds, still walk the earth! Do they really live on the fringes of the towering Himalayas and the edge of myth-haunted Tibet? From how many areas in the world have factual reports of wild, strange, hairy men emanated? Reports of strange apemen have come in from every continent, except Antarctica.

525 PAGES. 6x9 PAPERBACK. ILLUSTRATED. BIBLIOGRAPHY. INDEX. $16.95. CODE: ABML

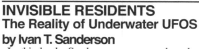

INVISIBLE RESIDENTS
The Reality of Underwater UFOS
by Ivan T. Sanderson

In this book, Sanderson, a renowned zoologist with a keen interest in the paranormal, puts forward the curious theory that "OINTS"—Other Intelligences—live under the Earth's oceans. This underwater, parallel, civilization may be twice as old as Homo sapiens, he proposes, and may have "developed what we call space flight." Sanderson postulates that the OINTS are behind many UFO sightings as well as the mysterious disappearances of aircraft and ships in the Bermuda Triangle. What better place to have an impenetrable base than deep within the oceans of the planet? Sanderson offers here an exhaustive study of USOs (Unidentified Submarine Objects) observed in nearly every part of the world.

298 PAGES. 6x9 PAPERBACK. ILLUSTRATED. BIBLIOGRAPHY. INDEX. $16.95. CODE: INVS

PIRATES & THE LOST TEMPLAR FLEET
The Secret Naval War Between the Templars & the Vatican
by David Hatcher Childress

Childress takes us into the fascinating world of maverick sea captains who were Knights Templar (and later Scottish Rite Free Masons) who battled the ships that sailed for the Pope. The lost Templar fleet was originally based at La Rochelle in southern France, but fled to the deep fiords of Scotland upon the dissolution of the Order by King Phillip. This banned fleet of ships was later commanded by the St. Clair family of Rosslyn Chapel (birthplace of Free Masonry). St. Clair and his Templars made a voyage to Canada in the year 1298 AD, nearly 100 years before Columbus! Later, this fleet of ships and new ones to come, flew the Skull and Crossbones, the symbol of the Knights Templar.

320 PAGES. 6x9 PAPERBACK. ILLUSTRATED. BIBLIOGRAPHY. $16.95. CODE: PLTF

TEMPLARS' LEGACY IN MONTREAL
The New Jerusalem
by Francine Bernier

The book reveals the links between Montreal and: John the Baptist as patron saint; Melchizedek, the first king-priest and a father figure to the Templars and the Essenes; Stella Maris, the Star of the Sea from Mount Carmel; the Phrygian goddess Cybele as the androgynous Mother of the Church; St. Blaise, the Armenian healer or "Therapeut"- the patron saint of the stonemasons and a major figure to the Benedictine Order and the Templars; the presence of two Black Virgins; an intriguing family coat of arms with twelve blue apples; and more.

352 PAGES. 6x9 PAPERBACK. ILLUSTRATED. BIBLIOGRAPHY. $21.95. CODE: TLIM

THE HISTORY OF THE KNIGHTS TEMPLARS
by Charles G. Addison, introduction by David Hatcher Childress

Chapters on the origin of the Templars, their popularity in Europe and their rivalry with the Knights of St. John, later to be known as the Knights of Malta. Detailed information on the activities of the Templars in the Holy Land, and the 1312 AD suppression of the Templars in France and other countries, which culminated in the execution of Jacques de Molay and the continuation of the Knights Templars in England and Scotland; the formation of the society of Knights Templars in London; and the rebuilding of the Temple in 1816. Plus a lengthy intro about the lost Templar fleet and its North American sea routes.

395 PAGES. 6x9 PAPERBACK. ILLUSTRATED. $16.95. CODE: HKT

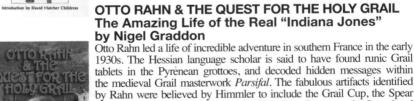

OTTO RAHN & THE QUEST FOR THE HOLY GRAIL
The Amazing Life of the Real "Indiana Jones"
by Nigel Graddon

Otto Rahn led a life of incredible adventure in southern France in the early 1930s. The Hessian language scholar is said to have found runic Grail tablets in the Pyrenean grottoes, and decoded hidden messages within the medieval Grail masterwork *Parsifal*. The fabulous artifacts identified by Rahn were believed by Himmler to include the Grail Cup, the Spear of Destiny, the Tablets of Moses, the Ark of the Covenant, the Sword and Harp of David, the Sacred Candelabra and the Golden Urn of Manna. Some believe that Rahn was a Nazi guru who wielded immense influence on his elders and "betters" within the Hitler regime, persuading them that the Grail was the Sacred Book of the Aryans, which, once obtained, would justify their extreme political theories and revivify the ancient Germanic myths. But things are never as they seem, and as new facts emerge about Otto Rahn a far more extraordinary story unfolds.

450 pages. 6x9 Paperback. Illustrated. $18.95. Code: ORQG

CASEBOOK ON THE MEN IN BLACK
By Jim Keith, Foreword by Kenn Thomas

UFO witnesses are sometimes intimidated by mysterious men dressed entirely in black. Are they government agents, sinister aliens or interdimensional creatures? Keith chronicles the strange goings on surrounding UFO activity and often bizarre cars that they arrive in—literal flying cars! Chapters include: Black Arts; Demons and Witches; Black Lodge; Maury Island; On a Bender; The Silence Group; Overlords and UMMO; More Black Ops; Indrid Cold; M.I.B.s in a Test Tube; Green Yard; The Hoaxers; Gray Areas; You Will Cease UFO Study; Beyond Reality; The Real/Unreal Men in Black; Deciphering a Nightmare; more. A new edition of the classic with updated material!

222 pages. 5x9 Paperback. Illustrated. $14.95. Code: CMIB

EYE OF THE PHOENIX
Mysterious Visions and
Secrets of the American Southwest
by Gary David

GaryDavid explores enigmas and anomalies in the vast American Southwest. Contents includes: The Great Pyramids of Arizona; Meteor Crater—Arizona's First Bonanza?; Chaco Canyon—Ancient City of the Dog Star; Phoenix—Masonic Metropolis in the Valley of the Sun; Along the 33rd Parallel—A Global Mystery Circle; The Flying Shields of the Hopi Katsinam; Is the Starchild a Hopi God?; The Ant People of Orion—Ancient Star Beings of the Hopi; Serpent Knights of the Round Temple; The Nagas—Origin of the Hopi Snake Clan?; The Tau (or T-shaped) Cross—Hopi/Maya/Egyptian Connections; The Hopi Stone Tablets of Techqua Ikachi; The Four Arms of Destiny—Swastikas in the Hopi World of the End Times; and more.

348 pages. 6x9 Paperback. Illustrated. $16.95. Code: EOPX

PRODIGAL GENIUS
The Life of Nikola Tesla
by John J. O'Neill

This special edition of O'Neill's book has many rare photographs of Tesla and his most advanced inventions. Tesla's eccentric personality gives his life story a strange romantic quality. He made his first million before he was forty, yet gave up his royalties in a gesture of friendship, and died almost in poverty. Tesla could see an invention in 3-D, from every angle, within his mind, before it was built; how he refused to accept the Nobel Prize; his friendships with Mark Twain, George Westinghouse and competition with Thomas Edison. Deluxe, illustrated edition.

408 pages. 6x9 Paperback. Illustrated. Bibliography.
$18.95. Code: PRG

STALKING THE TRICKSTERS:
Shapeshifters, Skinwalkers, Dark Adepts and 2012
By Christopher O'Brien
Foreword by David Perkins

Manifestations of the Trickster persona such as cryptids, elementals, werewolves, demons, vampires and dancing devils have permeated human experience since before the dawn of civilization. But today, very little is publicly known about The Tricksters. Who are they? What is their agenda? Known by many names including fools, sages, Loki, men-in-black, skinwalkers, shapeshifters, jokers, *jinn*, sorcerers, and witches, Tricksters provide us with a direct conduit to the unknown in the 21st century. Can these denizens of phenomenal events be attempting to communicate a warning to humanity in this uncertain age of prophesied change? Take a journey around the world stalking the tricksters!

354 Pages. 6x9 Paperback. Illustrated. Bibliography. $18.95. Code: STT

THE CRYSTAL SKULLS
Astonishing Portals to Man's Past
by David Hatcher Childress and Stephen S. Mehler

Childress introduces the technology and lore of crystals, and then plunges into the turbulent times of the Mexican Revolution form the backdrop for the rollicking adventures of Ambrose Bierce, the renowned journalist who went missing in the jungles in 1913, and F.A. Mitchell-Hedges, the notorious adventurer who emerged from the jungles with the most famous of the crystal skulls. Mehler shares his extensive knowledge of and experience with crystal skulls. Having been involved in the field since the 1980s, he has personally examined many of the most influential skulls, and has worked with the leaders in crystal skull research, including the inimitable Nick Nocerino, who developed a meticulous methodology for the purpose of examining the skulls.
294 pages. 6x9 Paperback. Illustrated. Bibliography. $18.95. Code: CRSK

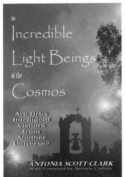

THE INCREDIBLE LIGHT BEINGS OF THE COSMOS
Are Orbs Intelligent Light Beings from the Cosmos?
by Antonia Scott-Clark

Scott-Clark has experienced orbs for many years, but started photographing them in earnest in the year 2000 when the "Light Beings" entered her life. She took these very seriously and set about privately researching orb occurrences. The incredible results of her findings are presented here, along with many of her spectacular photographs. With her friend, GoGos lead singer Belinda Carlisle, Antonia tells of her many adventures with orbs. Find the answers to questions such as: Can you see orbs with the naked eye?; Are orbs intelligent?; What are the Black Villages?; What is the connection between orbs and crop circles? Antonia gives detailed instruction on how to photograph orbs, and how to communicate with these Light Beings of the Cosmos.
334 pages. 6x9 Paperback. Illustrated. References. $19.95. Code: ILBC

AXIS OF THE WORLD
The Search for the Oldest American Civilization
by Igor Witkowski

Polish author Witkowski's research reveals remnants of a high civilization that was able to exert its influence on almost the entire planet, and did so with full consciousness. Sites around South America show that this was not just one of the places influenced by this culture, but a place where they built their crowning achievements. Easter Island, in the southeastern Pacific, constitutes one of them. The Rongo-Rongo language that developed there points westward to the Indus Valley. Taken together, the facts presented by Witkowski provide a fresh, new proof that an antediluvian, great civilization flourished several millennia ago.
220 pages. 6x9 Paperback. Illustrated. References. $18.95. Code: AXOW

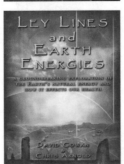

LEY LINE & EARTH ENERGIES
An Extraordinary Journey into the Earth's Natural Energy System
by David Cowan & Chris Arnold

The mysterious standing stones, burial grounds and stone circles that lace Europe, the British Isles and other areas have intrigued scientists, writers, artists and travellers through the centuries. How do ley lines work? How did our ancestors use Earth energy to map their sacred sites and burial grounds? How do ghosts and poltergeists interact with Earth energy? How can Earth spirals and black spots affect our health? This exploration shows how natural forces affect our behavior, how they can be used to enhance our health and well being.
368 PAGES. 6x9 PAPERBACK. ILLUSTRATED. $18.95. CODE: LLEE

SECRETS OF THE UNIFIED FIELD
The Philadelphia Experiment, the Nazi Bell, and the Discarded Theory
by Joseph P. Farrell

Farrell examines the now discarded Unified Field Theory. American and German wartime scientists and engineers determined that, while the theory was incomplete, it could nevertheless be engineered. Chapters include: The Meanings of "Torsion"; Wringing an Aluminum Can; The Mistake in Unified Field Theories and Their Discarding by Contemporary Physics; Three Routes to the Doomsday Weapon: Quantum Potential, Torsion, and Vortices; Tesla's Meeting with FDR; Arnold Sommerfeld and Electromagnetic Radar Stealth; Electromagnetic Phase Conjugations, Phase Conjugate Mirrors, and Templates; The Unified Field Theory, the Torsion Tensor, and Igor Witkowski's Idea of the Plasma Focus; tons more.

340 pages. 6x9 Paperback. Illustrated. Bibliography. Index. $18.95. Code: SOUF

NAZI INTERNATIONAL
The Nazi's Postwar Plan to Control Finance, Conflict, Physics and Space
by Joseph P. Farrell

Beginning with prewar corporate partnerships in the USA he moves on to the surrender of Nazi Germany, and evacuation plans of the Germans. He then covers the vast, and still-little-known recreation of Nazi Germany in South America with help of Juan Peron, I.G. Farben and Martin Bormann. Farrell then covers Nazi Germany's penetration of the Muslim world before moving on to the development and control of new energy technologies including the Bariloche Fusion Project, Dr. Philo Farnsworth's Plasmator, and the work of Dr. Nikolai Kozyrev. Finally, Farrell discusses the Nazi desire to control space, and examines their connection with NASA, the esoteric meaning of NASA Mission Patches.

412 pages. 6x9 Paperback. Illustrated. References. $19.95. Code: NZIN

ARKTOS
The Myth of the Pole in Science, Symbolism, and Nazi Survival
by Joscelyn Godwin

A scholarly treatment of catastrophes, ancient myths and the Nazi Occult beliefs. Explored are the many tales of an ancient race said to have lived in the Arctic regions, such as Thule and Hyperborea. Progressing onward, the book looks at modern polar legends including the survival of Hitler, German bases in Antarctica, UFOs, the hollow earth, Agartha and Shambala, more.

220 PAGES. 6x9 PAPERBACK. ILLUSTRATED. $16.95. CODE: ARK

GUARDIANS OF THE HOLY GRAIL
by Mark Amaru Pinkham

This book presents this extremely ancient Holy Grail lineage from Asia and how the Knights Templar were initiated into it. It also reveals how the ancient Asian wisdom regarding the Holy Grail became the foundation for the Holy Grail legends of the west while also serving as the bedrock of the European Secret Societies, which included the Freemasons, Rosicrucians, and the Illuminati. Also: The Fisher Kings; The Middle Eastern mystery schools, such as the Assassins and Yezidhi; The ancient Holy Grail lineage from Sri Lanka and the Templar Knights' initiation into it; The head of John the Baptist and its importance to the Templars; The secret Templar initiation with grotesque Baphomet, the infamous Head of Wisdom; more.

248 PAGES. 6x9 PAPERBACK. ILLUSTRATED. $16.95. CODE: GOHG

SCATTERED SKELETONS IN OUR CLOSET
By Karen Mutton
Australian researcher Mutton gives us the rundown on various hominids, skeletons, anomalous skulls and other "things" from our family tree, including hobbits, pygmies, giants and horned people. Chapters include: Human Origin Theories; Dating Techniques; Mechanisms of Darwinian Evolution; What Creationists Believe about Human Origins; Evolution Fakes and Mistakes; Creationist Hoaxes and Mistakes; The Tangled Tree of Evolution; The Australopithecine Debate; Homo Habilis; Homo Erectus; Anatomically Modern Humans in Ancient Strata?; Ancient Races of the Americas; Robust Australian Prehistoric Races; Pre Maori Races of New Zealand; The Taklamakan Mummies—Caucasians in Prehistoric China; Strange Skulls; Dolichocephaloids (Coneheads); Pumpkin Head, M Head, Horned Skulls; The Adena Skull; The Boskop Skulls; 'Starchild'; Pygmies of Ancient America; Pedro the Mountain Mummy; Hobbits—Homo Floresiensis; Palau Pygmies; Giants; Goliath; Holocaust of American Giants?; Giants from Around the World; more. Heavily illustrated.
320 Pages. 6x9 Paperback. Illustrated. $18.95. Code: SSIC

THE GRID OF THE GODS
The Aftermath of the Cosmic War and the Physics of the Pyramid Peoples
By Joseph P. Farrell with Scott D. de Hart
Physicist and Oxford-educated historian Farrell continues his best-selling book series on ancient planetary warfare, technology and the energy grid that surrounds the earth. Chapters on: Anomalies at the Temples of Angkor; The Ancient Prime Meridian: Giza; Transmitters, Temples, Sacred Sites and Nazis; Nazis and Geomancy; Nazi Transmitters and the Earth Grid; The Grid and Hitler's East Prussia Headquarters; Grid Geopolitical Geomancy; The Astronomical Correlation and the 10,500 BC Mystery; The Master Plan of a Hidden Elite; Moving and Immoveable Stones; Uncountable Stones and Stones of the Giants and Gods; Gateway Traditions; The Grid and the Ancient Elite; Finding the Center of the Land; The Ancient Catastrophe, the Very High Civilization, and the Post-Catastrophe Elite; The Meso- and South-American "Pyramid Peoples"; Tiahuanaco and the Puma Punkhu Paradox: Ancient Machining; The Mayans, Their Myths and the Mounds; The Pythagorean and Platonic Principles of Sumer, Babylonia and Greece; The Gears of Giza: the Center of the Machine; Alchemical Cosmology and Quantum Mechanics in Stone: The Mysterious Megalith of Nabta Playa; The Physics of the "Pyramid Peoples"; tons more.
436 Pages. 6x9 Paperback. Illustrated. References. $19.95. Code: GOG

BEYOND EINSTEIN'S UNIFIED FIELD
Gravity and Electro-Magnetism Redefined
By John Brandenburg, Ph.D.
Veteran plasma physicist John Brandenburg shows the intricate interweaving of Einstein's work with that of other physicists, including Sarkharov and his "zero point" theory of gravity and the hidden fifth dimension of Kaluza and Klein. He also traces the surprising, hidden influence of Nikola Tesla on Einstein's life. Brandenburg describes control of space-time geometry through electromagnetism, and states that faster-than-light travel will be possible in the future. Anti-gravity through electromagnetism is possible, which upholds the basic "flying saucer" design utilizing "The Tesla Vortex." See the physics used at Area 51 explained! Chapters include: Squaring the Circle, Einstein's Final Triumph; Mars Hill, or the Cosmos As It Is; A Book of Numbers and Forms; Kepler, Newton and the Sun King; Magnus and Electra; Atoms of Light; Einstein's Glory, Relativity; The Aurora; Tesla's Vortex and the Cliffs of Zeno; The Hidden 5th Dimension; The GEM Unification Theory; Anti-Gravity and Human Flight; The New GEM Cosmos; Summit of Mount Einstein; more. Includes and 8-page color section.
312 Pages. 6x9 Paperback. Illustrated. References. $18.95. Code: BEUF

ORDER FORM

10% Discount When You Order 3 or More Items!

One Adventure Place
P.O. Box 74
Kempton, Illinois 60946
United States of America
Tel.: 815-253-6390 • Fax: 815-253-6300
Email: auphq@frontiernet.net
http://www.adventuresunlimitedpress.com

ORDERING INSTRUCTIONS

✓ Remit by USD$ Check, Money Order or Credit Card

✓ Visa, Master Card, Discover & AmEx Accepted

✓ Paypal Payments Can Be Made To:
   info@wexclub.com

✓ Prices May Change Without Notice

✓ 10% Discount for 3 or more Items

SHIPPING CHARGES

United States

✓ Postal Book Rate { $4.00 First Item / 50¢ Each Additional Item

✓ POSTAL BOOK RATE Cannot Be Tracked!

✓ Priority Mail { $5.00 First Item / $2.00 Each Additional Item

✓ UPS { $6.00 First Item / $1.50 Each Additional Item

   NOTE: UPS Delivery Available to Mainland USA Only

Canada

✓ Postal Air Mail { $10.00 First Item / $2.50 Each Additional Item

✓ Personal Checks or Bank Drafts MUST BE
   US$ and Drawn on a US Bank

✓ Canadian Postal Money Orders OK

✓ Payment MUST BE US$

All Other Countries

✓ Sorry, No Surface Delivery!

✓ Postal Air Mail { $16.00 First Item / $6.00 Each Additional Item

✓ Checks and Money Orders MUST BE US$
   and Drawn on a US Bank or branch.

✓ Paypal Payments Can Be Made in US$ To:
   info@wexclub.com

SPECIAL NOTES

✓ RETAILERS: Standard Discounts Available

✓ BACKORDERS: We Backorder all Out-of-Stock Items Unless Otherwise Requested

✓ PRO FORMA INVOICES: Available on Request

ORDER ONLINE AT: www.adventuresunlimitedpress.com

Please check: ☑

☐ This is my first order   ☐ I have ordered before

Name

Address

City

State/Province _____ Postal Code

Country

Phone day _____ Evening

Fax _____ Email

Item Code	Item Description	Qty	Total

Please check: ☑

	Subtotal ▶	
	Less Discount-10% for 3 or more items ▶	
☐ Postal-Surface	Balance ▶	
☐ Postal-Air Mail (Priority in USA)	Illinois Residents 6.25% Sales Tax ▶	
	Previous Credit ▶	
☐ UPS	Shipping ▶	
(Mainland USA only)	Total (check/MO in USD$ only) ▶	

☐ Visa/MasterCard/Discover/American Express

Card Number

Expiration Date

10% Discount When You Order 3 or More Items!